THE RELIABILITY FRONTIER

How America's Grid Is Racing to Meet the AI Era

Traditional Power, Renewables, Batteries, and the Fight to Keep the Lights On

Douglas Donovan

IBSN 979-8-2562-2303-8

Preface

For most of its history, the American electric grid was the most boring miracle in the country. Flip a switch, and light appears. Plug in a refrigerator, and food stays cold for decades without a second thought. Generations of Americans grew up trusting that the grid would simply work, and for the most part, it did. Reliability was not a subject of dinner-table conversation; it was a background assumption so deeply embedded in daily life that most people could not name their local utility, much less explain how power actually reached their homes.

That quiet confidence has been cracking for several years, and the cracks are now impossible to ignore. Texans shivered in the dark during Winter Storm Uri in February 2021, watching pipes burst and neighbors die of carbon monoxide poisoning while running generators indoors for warmth. Californians grew accustomed to rolling blackouts during heat waves and wildfire-driven public safety power shutoffs. Grid operators across the country began issuing more frequent conservation appeals, asking households to set thermostats a few degrees warmer in August afternoons to avoid emergency measures. The North American Electric Reliability Corporation — the industry's own watchdog — began publishing annual reliability assessments with words like "elevated risk" and "high risk" attached to larger and larger swaths of the country's map.

Then came a new and unexpected pressure, one that had nothing to do with storms, wildfires, or aging power plants: artificial intelligence. Beginning around 2022 and accelerating sharply through 2023, 2024, and 2025, hyperscale technology companies — Microsoft, Amazon, Google, Meta, and a growing roster of specialized AI infrastructure firms — began proposing datacenters of a size and power intensity the grid had never been asked to serve. A single AI training campus can request as much electricity as a mid-sized city. Utilities that had spent a decade planning for flat or slowly declining electricity demand were suddenly presented with interconnection requests that, if built, would require doubling or tripling local generation and transmission capacity within a matter of years, not decades.

This book is an attempt to explain, in plain language, how these forces intersect: the grid's traditional reliance on coal, natural gas, nuclear, and hydroelectric generation; the rapid rise of solar and wind power and the operational challenges that intermittent resources create; the equally rapid emergence of utility-scale battery storage as a tool for smoothing out those challenges; the demand shock created by AI datacenters and the strategies utilities are deploying to meet it; the quiet but consequential shift in how grid operators handle capacity emergencies, including the increasing likelihood that large, flexible customers like datacenters will be curtailed before residential neighborhoods lose power; the emerging promise of small modular nuclear reactors and other next-generation generation sources; and finally, how all of this compares to the extraordinary pace of generation capacity additions underway in China.

None of these threads exists in isolation. The grid is a single, interconnected physical system, and every decision made in one domain — a coal plant retirement, a battery installation, a datacenter interconnection agreement, a transmission line permit — ripples through all the others. Understanding American electric reliability in the 2020s and 2030s requires holding all of these threads at once. That is the task this book sets for itself.

I have written this book for the general reader and the engaged undergraduate: someone who reads the news about grid strain, AI energy demand, or blackout risk and wants a deeper, structured understanding without needing an engineering degree to follow along. Where technical concepts are unavoidable — capacity markets, frequency regulation, grid-forming inverters — I have tried to explain them from first principles, building the vocabulary needed for later chapters as we go.

The grid is entering its most consequential period of change since rural electrification brought power to nearly every American home in the mid-twentieth century. This book is an attempt to make sense of that change while it is still happening.

Introduction:

A Grid Built for a Different Century

Picture the American electric grid as it existed in roughly 1970. Power plants were large, centralized, and almost entirely fueled by coal, oil, natural gas, hydroelectric dams, and a rapidly expanding fleet of nuclear reactors. Electricity flowed in one direction: from the power plant, through high-voltage transmission lines, down to lower-voltage distribution lines, and finally into homes and businesses. Demand was remarkably predictable — it rose steadily year after year in a smooth, almost boring curve that utility planners could forecast with confidence a decade or more into the future. Utilities built generation to match that forecast, regulators approved the costs, and customers paid for it through steadily rising but predictable rates. This was the grid's first great architecture: centralized, one-directional, and fossil-and-nuclear-anchored.

That architecture served the country well for half a century, but it was designed to solve a specific problem — delivering ever-increasing amounts of electricity from a small number of large power plants to a growing but predictable set of customers. It was not designed to solve the problems the grid faces today: integrating enormous quantities of weather-dependent renewable generation, accommodating power flows that increasingly move in multiple directions as homes and businesses generate their own electricity, and absorbing sudden, enormous, geographically concentrated new loads that can appear on a two-to-four-year timeline rather than a multi-decade one.

Three forces are converging to reshape this architecture simultaneously, and understanding each of them — and how they interact — is the purpose of this book.

The first force is decarbonization. Driven by state renewable portfolio standards, corporate sustainability commitments, falling technology costs, and federal incentives such as the Inflation Reduction Act's clean energy tax credits, the United States has added solar and wind generation at a pace that would have seemed unimaginable twenty years ago. Utility-scale solar capacity has grown from a rounding error in the national generation mix to a resource that, in states like California and Texas, provides a substantial share of daytime electricity. Wind power, concentrated in the Great Plains corridor stretching from Texas to the Dakotas and increasingly along the Atlantic coast, has become one of the largest sources of new generation capacity added to the grid each year. This transition brings real benefits — lower fuel costs, reduced emissions, and diversification away from commodity price volatility — but it also brings a fundamental operational challenge: the sun does not always shine and the wind does not always blow, and a grid built around large, controllable, "dispatchable" power plants must be re-engineered to handle resources that are only available on nature's schedule.

The second force is the retirement of traditional generation, running on a collision course with the first. Coal plants, facing economic pressure from cheap natural gas and renewable energy alongside environmental regulation, have been retiring across the country for over a decade. Aging nuclear plants have faced their own economic pressures, though a wave of subsidies, tax credits, and — as this book will discuss — new corporate demand from AI datacenter operators has begun to reverse some planned closures. Reliability regulators have grown increasingly vocal about the risk that generation retirements are outpacing the buildout of adequate replacement capacity, particularly resources that can be relied upon at any hour of the day, in any season, regardless of weather.

The third force, and the newest, is an explosion in electricity demand driven substantially by artificial intelligence infrastructure. For roughly fifteen years, from the mid-2000s through the early 2020s, aggregate U.S. electricity demand was essentially flat. Energy efficiency improvements in lighting, appliances, and industrial processes largely offset population and economic growth. Utility long-term planning departments grew accustomed to modest, low-single-digit-percent annual load growth forecasts, if any growth at all. That era has ended abruptly. Data centers built to train and run large AI models consume electricity at a scale and with a load profile — enormous, constant, and fast-ramping — that utilities have not had to plan around before. Combined with the electrification of transportation and heating, and a resurgence in domestic manufacturing, the United States is now facing the fastest sustained electricity demand growth forecast in a generation, concentrated heavily in a handful of regions.

These three forces do not operate independently. A grid that is simultaneously retiring dispatchable generation, adding large quantities of weather-dependent generation, and absorbing an unprecedented surge in new demand faces a reliability challenge unlike anything in its history. Utility-scale battery storage has emerged as perhaps the single most important new tool for managing this challenge, capable of absorbing surplus renewable generation during the day and discharging it during the evening demand peak, while also providing the split-second frequency stabilization services that spinning fossil and nuclear generators have traditionally supplied "for free" as a byproduct of their mechanical operation.

Meanwhile, grid operators are quietly rewriting the rules of last resort. For decades, the sequence of events during a true capacity emergency was well understood: reduce voltage, appeal to the public for conservation, and if necessary, implement rotating outages that would affect residential and industrial customers roughly in proportion to their share of system load. That sequence is being reordered. Utilities are increasingly negotiating contracts with their largest customers — foremost among them AI datacenters, many of which already maintain substantial on-site backup generation for their own reliability purposes — that allow the grid operator to curtail those customers first, before residential neighborhoods or critical industrial facilities lose power. This is not a hypothetical policy proposal; it is being actively implemented through interconnection tariffs, demand response contracts, and market rule changes across multiple regions.

Looking further ahead, a new generation of nuclear technology — small modular reactors, built in factories and assembled on site rather than constructed as one-of-a-kind mega-projects — promises to offer a source of firm, carbon-free generation that could be co-located directly with the datacenters and industrial facilities driving demand growth. Several major technology companies have already signed agreements to develop or restart nuclear capacity specifically to power their AI ambitions. Whether this technology can be deployed at the pace and scale required remains one of the central open questions this book will examine.

Finally, no honest account of American grid reliability can ignore the international context. China is adding electricity generation capacity — coal, solar, wind, nuclear, and ultra-high-voltage transmission alike — at a pace that dwarfs anything happening in the United States, executing in a single year what American utilities might take a decade to permit and build. Understanding whether the United States can meet its own future load growth requires understanding how that growth compares to what other major economies are doing, and what advantages and disadvantages the American system's market-based, locally regulated structure creates relative to China's state-directed model.

This book proceeds in six parts. Part I establishes the foundational architecture of the American grid and its continued reliance on traditional generation sources. Part II examines the rapid rise of solar and wind power and the emergence of battery storage as the resource that makes higher renewable penetration operationally possible. Part III turns to the demand shock: how AI datacenters are reshaping load forecasts, what utilities are doing to meet that demand, and how the rules governing emergency load shedding are being rewritten to protect residential and critical customers first. Part IV looks to the future of generation technology, particularly small modular reactors. Part V places the American experience in global context by examining China's extraordinary generation buildout. Part VI synthesizes these threads and looks ahead to what grid reliability is likely to mean by the mid-2030s.

The grid that emerges from this period of transformation will look very different from the grid of 1970, or even 2010. Whether it is more reliable, more resilient, and better prepared for the demands of an AI-powered economy depends on choices being made right now, in utility planning departments, state regulatory commissions, federal energy offices, and corporate boardrooms across the country. This book is an attempt to help the reader understand those choices and their consequences.

How to Read This Book

This book is organized so that each Part builds on the concepts and vocabulary established in the Parts before it, and readers new to grid and energy policy topics will likely benefit most from reading it in order, particularly Part I, which establishes the foundational vocabulary — balancing authorities, reserve margins, capacity markets, and ancillary services — that every subsequent chapter relies upon without re-explaining. Readers who already possess a working knowledge of how the grid operates, perhaps from professional experience in the utility or energy policy sector, may reasonably skip ahead to Part III, where this book's most novel arguments about AI datacenter demand growth and the reordering of emergency load-shedding priorities are developed, returning to Parts I and II as needed for specific technical grounding on solar, wind, and storage.

Each chapter is written to stand reasonably well on its own, with cross-references to related discussions in other chapters clearly noted, so that readers using this book as a reference — returning to a specific chapter to refresh their understanding of small modular reactors, for instance, or the specifics of PJM's capacity market — should be able to do so without needing to re-read the entire book from the beginning. The glossary in the back matter provides a quick reference for the technical terminology introduced throughout the book, and the timeline of major grid events offers a chronological anchor for readers seeking to place the developments discussed throughout the book in sequence.

A note on figures and forecasts: as discussed at greater length in Chapter 14's concluding discussion of forecasting uncertainty, this book was written during a period of genuinely rapid change in AI datacenter demand forecasts, capacity market prices, and individual project timelines. Where this book cites specific figures — a percentage growth forecast, a capacity market clearing price increase, a project completion date — these figures reflect the best publicly available information at the time of writing and are offered to illustrate the scale and direction of the trends this book describes, not as precise, unchanging facts that readers should expect to remain accurate indefinitely. Readers seeking the most current version of any specific figure are encouraged to consult the primary sources described in the Notes on Sources section of the back matter.

A Word on Scope

This book focuses specifically on the intersection of traditional generation, renewable energy, battery storage, AI-driven demand growth, emergency load-shedding protocols, emerging nuclear and other generation technologies, and the U.S.-China infrastructure comparison. It is not intended as a comprehensive treatise on every dimension of American energy policy, electricity markets, or climate policy; readers interested in retail rate design, broader climate policy debates, grid cybersecurity, or the deeper technical engineering of power system operation will find this book a useful entry point and contextual grounding, but should seek out the more specialized literature that exists on each of those adjacent topics for a fuller treatment. What this book aims to offer instead is a single, coherent narrative connecting several major forces reshaping American grid reliability that are too often discussed separately — in one news article about datacenter electricity demand, in another about battery storage costs, in a third about capacity market prices — when in fact, as this book has aimed to demonstrate throughout its fourteen chapters, they are best understood as a single, interconnected story.

PART I

THE GRID WE INHERITED

Chapter 1: How the Grid Actually Works

To understand why AI datacenters, battery storage, and small modular reactors matter to grid reliability, it is first necessary to understand what the grid is and how it maintains the invisible balancing act that keeps electricity flowing twenty-four hours a day. This chapter builds the vocabulary and conceptual foundation used throughout the rest of the book.

Generation, Transmission, and Distribution

The electric power system is conventionally divided into three physical layers.

Generation is where electricity is produced — at coal, natural gas, nuclear, and hydroelectric power plants, at wind farms and solar arrays, and increasingly at battery storage facilities that, while they do not create electricity from a primary fuel source, function operationally as generation resources by injecting stored power back onto the grid when needed.

Transmission is the high-voltage highway system that moves large quantities of electricity over long distances, typically at voltages between 138,000 and 765,000 volts. Transmission lines are the tall steel lattice towers most people picture when they think of "power lines" crossing open country. High voltages are used for long-distance transmission because they dramatically reduce the energy lost as heat during transport — a critical fact, since a shortage of transmission capacity between where cheap renewable power is generated and where demand is concentrated is one of the central bottlenecks discussed later in this book.

Distribution is the local network that steps voltage back down through substations and neighborhood transformers, delivering electricity at the 120- and 240-volt levels used in homes and businesses. Distribution networks are typically owned and operated by the same utility that sells retail electricity to a given community, while transmission networks are increasingly operated by regional entities that span multiple states and multiple utility service territories.

Balancing Authorities and the Real-Time Matching Problem

Unlike most commodities, electricity cannot be economically stored in bulk at the scale of an entire grid — or at least, this was true until the recent emergence of utility-scale batteries, a development significant enough that Chapter 5 is devoted to it. For most of the grid's history, this meant that supply and demand had to be matched instantaneously, every second of every day, across an enormous interconnected network. If demand exceeds available supply even briefly, the grid's frequency — nominally 60 cycles per second (hertz) in North America — begins to drop. If supply exceeds demand, frequency rises. Deviations of even a fraction of a hertz, sustained for more than a few seconds, can damage generating equipment and, in extreme cases, trigger automatic protective systems that disconnect power plants or entire sections of the grid to prevent cascading failure — the kind of event that produced the Northeast blackout of 2003, which left fifty million people without power.

The entities responsible for managing this real-time balance are called **balancing authorities**. A balancing authority continuously monitors electricity supply and demand within its footprint and dispatches generation resources up or down in real time to keep the system in balance. In much of the United States, balancing authority functions have been consolidated into large regional organizations known as **Independent System Operators (ISOs)** or **Regional Transmission Organizations (RTOs)**. These entities do not typically own power plants or transmission lines themselves; instead, they operate wholesale electricity markets, dispatch generation from many different owners according to economic and reliability rules, and plan the transmission expansions needed to keep the system reliable.

The major ISOs and RTOs relevant to this book include:

- **PJM Interconnection**, covering all or part of thirteen states and the District of Columbia, from Illinois and Michigan through Pennsylvania, Ohio, Virginia, and into the Carolinas. PJM is the largest wholesale electricity market in the world by footprint and is especially important to this book's story because it covers Northern Virginia — the single largest concentration of datacenter load on the planet, sometimes called "Data Center Alley."
- **ERCOT (Electric Reliability Council of Texas)**, which operates the electricity grid for roughly 90 percent of Texas. ERCOT is unique among U.S. grid operators in that it is almost entirely electrically isolated from the rest of the country, a deliberate historical choice made to avoid federal regulatory jurisdiction. This isolation was a central factor in the severity of the February 2021 Winter Storm Uri crisis, discussed in the next chapter, because ERCOT could not import large quantities of emergency power from neighboring regions the way an interconnected grid operator could.
- **MISO (Midcontinent Independent System Operator)**, spanning a north-south corridor from Manitoba, Canada, through the upper Midwest and down into Louisiana and Mississippi. MISO's footprint includes some of the best wind resources in North America.
- **CAISO (California Independent System Operator)**, which operates most of California's grid and has been at the leading edge of high renewable energy penetration, along with the

operational challenges — most famously the "duck curve," discussed in Chapter 3 — that come with it.

- **SPP (Southwest Power Pool)**, covering much of the central Great Plains, another wind-rich region.
- The **Southeast and much of the Western United States** outside California, where vertically integrated utilities (utilities that own generation, transmission, and distribution together, such as Georgia Power, Duke Energy, and Dominion Energy) continue to perform balancing authority functions themselves or through smaller regional arrangements, without a large multi-state ISO/RTO market overlay. This distinction matters throughout this book because the Southeast — with Georgia, Virginia, and the Carolinas hosting enormous datacenter growth — operates under a different planning and regulatory model than the ISO/RTO regions, with more centralized utility control over generation decisions and a correspondingly different set of tools for managing large new loads.

Reserve Margins and Capacity Adequacy

Because electricity demand fluctuates by time of day, day of week, and season — and because power plants can unexpectedly fail or need to be taken offline for maintenance — grid planners do not build generation capacity to exactly match expected peak demand. Instead, they maintain a **reserve margin**: extra generating capacity beyond the forecasted peak, expressed as a percentage. A reserve margin of 15 percent means a system has generating capacity equal to 115 percent of its expected peak demand.

Reserve margins are not arbitrary; they are calculated using probabilistic reliability standards, most commonly a target called "one day in ten years" loss-of-load expectation, meaning the system is designed so that, on average, involuntary firm load shedding due to a resource shortfall should occur no more than once per decade. NERC, the

North American Electric Reliability Corporation, is the federally designated organization responsible for developing and enforcing mandatory reliability standards across the U.S. and Canadian bulk power system, and it publishes an annual Long-Term Reliability Assessment that has, in recent years, become essential reading for anyone following the story told in this book. NERC's assessments have grown steadily more alarmed, identifying larger portions of North America at "elevated" or "high" risk of resource shortfalls during extreme conditions over the coming decade — a direct consequence of the three forces described in this book's introduction.

Capacity Markets

In ISO/RTO regions, reserve margins are not simply mandated by fiat; they are substantially procured through **capacity markets**, a mechanism designed to ensure that enough generation (and, increasingly, demand response and storage) will be available in future years to meet reliability targets. PJM's capacity market, known as the Reliability Pricing Model, conducts auctions in which generation owners bid to commit their plants to be available during a future delivery year in exchange for a capacity payment — revenue distinct from the money they earn actually selling electricity. If insufficient capacity is offered at prices generators find acceptable, the auction clears at a higher price, sending an economic signal that more capacity is needed.

These capacity auctions have become one of the most closely watched barometers of the reliability stress described throughout this book. PJM's capacity auction for the 2025/2026 delivery year cleared at prices roughly nine times higher than the prior year's auction, a dramatic increase that PJM and independent market monitors attributed substantially to a combination of generator retirements, slower-than-needed new resource additions, and — critically — surging load growth forecasts driven by datacenter demand. That single data point is, in many ways, the clearest quantitative signal available that the demand shock described in Part III of this book is already reshaping electricity markets, not merely appearing in speculative future forecasts.

Frequency Regulation and Ancillary Services

Beyond simply having enough total generation capacity, grid operators must maintain the finer-grained, second-by-second services that keep the grid's frequency stable. These are collectively known as **ancillary services** and include:

- **Frequency regulation**, the continuous, automatic adjustment of generation output to counteract small, constant fluctuations in supply and demand.
- **Spinning reserve**, generation capacity that is already online and synchronized to the grid but not fully loaded, able to increase output within seconds to minutes if another resource unexpectedly trips offline.
- **Non-spinning (or supplemental) reserve**, additional capacity that can start up and reach full output within a slightly longer window, typically ten to thirty minutes.
- **Black start capability**, the ability of certain generating units to restart without external power, essential for recovering from a total or partial blackout.
- **Inertia**, a more subtle but increasingly important concept discussed further in Chapter 10: the physical, mechanical resistance to sudden frequency change provided by the spinning mass of traditional turbine generators. As more of the grid's generation comes from inverter-based resources like solar and wind, which do not inherently provide this mechanical inertia, grid operators have had to develop new tools — including "grid-forming" battery inverters — to replace a service the system used to receive automatically.

Why This Matters for Everything That Follows

Every chapter in this book connects back to the concepts introduced here. When Chapter 3 discusses solar's "duck curve," it is describing a reserve margin and ramping challenge. When Chapter 5 discusses batteries providing frequency regulation, it is describing an ancillary service that used to come primarily from fossil generators. When Chapter 6 discusses AI datacenters reshaping load forecasts, it is describing a fundamental input to the reserve margin and capacity market calculations described above. When Chapter 8 discusses curtailing datacenters before residential customers during an emergency, it is describing a modification to the load-shedding protocols that exist precisely because reserve margins can, in extreme conditions, be exceeded despite the best planning. And when Chapter 9 discusses small modular reactors, it is describing a potential new source of the firm, dispatchable, inertia-providing capacity that the system has historically depended upon.

The grid, in other words, is a single integrated machine, and every actor within it — power plant owner, transmission planner, ISO market designer, datacenter developer, residential customer — is a participant in the same continuous balancing act. The remainder of this book examines how that balancing act is being reshaped by the forces of decarbonization, generation retirement, and unprecedented demand growth, and what it will take to keep the lights on through the transition.

The Physical Reality Behind the Abstractions

It is easy to discuss balancing authorities, capacity markets, and ancillary services as though they were purely financial or administrative constructs, but it is worth pausing to appreciate the physical reality underlying them. At any given moment, the electricity flowing into a home in Ohio was very likely generated somewhere within the past fraction of a second — not stored, not shipped, not warehoused, but produced essentially simultaneously with its consumption, having traveled at close to the speed of light through a network of wires, transformers, and switching stations that together represent one of the largest and most complex machines ever built by human beings. The entire North American grid, excluding Texas's largely isolated ERCOT system and a handful of smaller exceptions, operates as two enormous synchronized alternating-current systems — the Eastern and Western Interconnections — within which every generator, from a household rooftop solar panel to a two-thousand-megawatt nuclear plant, must remain in near-perfect frequency and phase synchronization with every other generator, all day, every day, indefinitely. Losing that synchronization, even briefly, in a large enough section of the system can cascade into the kind of wide-area blackout that struck the Northeast in August 2003, when a combination of a software bug, untrimmed trees contacting transmission lines in Ohio, and a series of cascading equipment trips left fifty million people across the northeastern United States and Ontario without power, some for as long as four days.

This physical reality is why grid reliability cannot be treated as a purely financial or regulatory abstraction, and why this book spends so much time on the specific engineering trade-offs — inertia, ramping capability, fuel supply chains — that determine whether the lights actually stay on. Markets and regulations exist to create incentives that, ideally, produce a physical system with the attributes reliability requires; but the physical requirements come first, and no amount of clever market design can substitute for having enough of the right kind of generation, transmission, and storage physically built, interconnected, and available when needed.

The Historical Arc from Local Utilities to Regional Markets

The regional ISO/RTO structure described in this chapter did not always exist. For most of the twentieth century, American electric utilities were vertically integrated, geographically distinct monopolies, each responsible for generation, transmission, and distribution within its own service territory, with relatively limited coordination beyond emergency mutual assistance agreements and modest power trading among neighboring utilities. This structure began to change in the 1990s and 2000s, driven by a wave of electricity market deregulation and restructuring at both the federal and state level, motivated by a belief — drawn substantially from economic theory suggesting that competitive markets would produce more efficient outcomes than regulated monopolies — that opening wholesale electricity generation to competition would lower costs and improve efficiency compared to the traditional vertically integrated utility model.

The Federal Energy Regulatory Commission's Order 888 in 1996 and Order 2000 in 1999 encouraged the formation of the regional transmission organizations described in this chapter, aiming to create larger, more efficient wholesale markets and to separate transmission operation from the generation-owning utilities that might otherwise have an incentive to favor their own power plants over competitors seeking to use the same transmission lines. Not all states embraced this restructuring; the Southeast, much of the Mountain West, and other regions retained the traditional vertically integrated utility model, in which state public utility commissions directly regulate the rates and investment decisions of utilities that continue to own generation, transmission, and distribution together. This split — roughly two-thirds of the country operating under ISO/RTO wholesale markets, with the remainder under traditional vertically integrated regulation — persists today and matters enormously for this book's later chapters, because it means the tools available to serve a large new datacenter customer, and the regulatory process for approving large-load tariffs and curtailment provisions, differ meaningfully depending on which of these two regulatory models governs a given region.

Measuring Reliability: SAIDI, SAIFI, and Beyond

Utilities and regulators track grid reliability using a standard set of metrics that, while less prominent in public discussion than dramatic events like Winter Storm Uri, provide the ongoing, granular data that utility planners and regulators actually use to evaluate system performance day to day. The **System Average Interruption Duration Index (SAIDI)** measures the average total number of minutes a typical customer experiences an outage over a given year, while the **System Average Interruption Frequency Index (SAIFI)** measures the average number of separate outage events a typical customer experiences over the same period. These metrics are typically reported both including and excluding "major event days" — days affected by severe storms or other extraordinary circumstances — since a small number of severe weather events can otherwise dominate and distort a utility's annual reliability statistics relative to its more routine day-to-day performance.

These distribution-level reliability metrics are conceptually distinct from, though related to, the bulk-system resource adequacy concerns — reserve margins, capacity markets, and the risk of involuntary load shedding — that dominate this book's discussion. The overwhelming majority of the outages captured in SAIDI and SAIFI statistics stem from localized distribution system events: a tree falling on a line, a vehicle striking a utility pole, equipment failure at a neighborhood transformer, rather than from the kind of system-wide generation shortfall this book focuses on. Distribution system reliability remains an important, ongoing utility responsibility in its own right, but it is largely a separate engineering and investment challenge from the bulk-system resource adequacy questions — will there be enough generation and transmission capacity to meet demand across an entire region — that form this book's central focus, and readers should keep this distinction in mind when encountering reliability statistics in utility filings or news coverage, since the two categories of metrics measure genuinely different things and are not always clearly distinguished in public discussion.

Frequency Response In Practice: A Walk Through a Contingency Event

To make the abstract concept of frequency regulation more concrete, consider what actually happens, second by second, when a large power plant unexpectedly trips offline — a scenario grid operators call a "contingency event" and plan around constantly. In the first fraction of a second after the plant trips, the collective inertia of every other synchronous generator still spinning on the interconnected system begins to slow the resulting frequency decline, buying precious time. Within seconds, automatic governor response — built into the control systems of many conventional generators and, increasingly, into grid-forming battery inverters — begins nudging other resources' output upward to help arrest the decline. Within roughly ten seconds to a few minutes, primary frequency response and fast-acting reserves, including battery storage, come online more fully. Within ten to thirty minutes, slower-starting reserves and, if necessary, emergency energy purchases from neighboring balancing authorities are deployed to fully restore balance and allow the faster reserves to stand back down and remain available for the next contingency.

This layered, time-phased response architecture — inertia, then automatic governor response, then fast reserves, then slower reserves, then market-based rebalancing — is designed so that a single contingency, even a very large one, such as the sudden loss of the single largest generator or transmission line on the system (a standard grid operators call the "N-1" criterion, ensuring the system can withstand the loss of any single element without cascading failure), does not cause frequency to deviate far enough to trigger automatic protective relays that would disconnect additional equipment and risk a cascading, self-reinforcing blackout. Understanding this layered architecture helps explain why the grid-forming inverter technology discussed in Chapter 10, and the fast-responding battery storage discussed in Chapter 5, have become such a significant focus of engineering attention as more of the grid's spinning, inertia-providing traditional generation is displaced by inverter-based renewable and storage resources: each layer of this response architecture must continue functioning effectively even as the underlying mix of resources providing it changes fundamentally.

Why Regional Diversity Itself Is a Reliability Asset

A final foundational concept worth introducing in this opening chapter, because it recurs throughout the remainder of the book, is the reliability value of sheer geographic and resource diversity. A weather event severe enough to threaten reliability in one region — a heat dome over Texas, a cold snap across the Midwest — is rarely simultaneously affecting every region of the country with equal severity, meaning that interconnections between regions, allowing power to flow from an unaffected region to a stressed one, provide a form of reliability insurance that no single region can replicate on its own, regardless of how much generation capacity it builds within its own borders. This is precisely why ERCOT's near-total electrical isolation from neighboring grids, noted earlier in this chapter, has been so consequential to Texas's reliability challenges: unlike a utility within the interconnected Eastern or Western grid, which can lean on emergency imports from neighboring regions during a localized crisis, ERCOT has historically had only limited direct-current interconnection ties to neighboring grids, capable of importing only a modest amount of emergency power relative to the scale of a crisis like Winter Storm Uri. This geographic-diversity principle will resurface throughout this book, including in the discussion of interregional transmission's reliability value in Chapter 4, and in the comparison to China's nationally coordinated transmission planning in Part V.

The Vocabulary Carried Forward

The concepts introduced in this chapter — balancing authorities and the ISO/RTO landscape, reserve margins and the one day-in-ten-years reliability standard, capacity markets and their price signals, and the layered ancillary services architecture that keeps grid frequency stable second by second — form the vocabulary this book relies upon throughout every subsequent chapter. When Chapter 6 describes AI datacenters straining utility load forecasts, it is describing a challenge to the reserve margin and capacity market framework introduced here. When Chapter 8 describes the reordering of emergency load-shedding priorities, it is describing a modification to the same emergency operating procedures whose escalating sequence this chapter has outlined in brief. Readers who internalize this chapter's foundational concepts will find the more novel, specific arguments of later chapters considerably easier to follow, evaluate, and ultimately apply to their own understanding of the American grid's ongoing transformation.

Chapter 2:

The Traditional Backbone — Coal, Gas, Nuclear, Hydro

Despite two decades of rapid renewable energy growth, the American grid in the mid-2020s still depends overwhelmingly on traditional generation sources for the majority of its electricity, and even more overwhelmingly for its reliability during the hardest hours of the year – the coldest winter mornings and the hottest summer evenings, when demand peaks and renewable output is often at its least helpful. This chapter examines why traditional generation remains the backbone of the system, the pressures pushing pieces of that backbone into retirement, and what happens when the backbone is stressed past its limits.

Coal: The Long Goodbye

Coal-fired power plants once supplied more than half of all U.S. electricity. That share has fallen to roughly 15 percent nationally as of the mid-2020s, and it continues to decline. The reasons are primarily economic rather than purely regulatory: the shale gas revolution of the 2010s made natural gas so cheap that it became more economical to build and run gas plants than to continue operating, let alone build new, coal plants. Environmental regulations – governing everything from mercury and air toxics to coal ash disposal to, more recently, greenhouse gas emissions – have added compliance costs that further tilted the economics against coal. Meanwhile, wind and solar power, once heavily subsidized novelties, became by the early 2020s the cheapest source of new electricity generation in much of the country on a pure cost-per-megawatt-hour basis.

The result has been a wave of coal plant retirements running through the 2010s and 2020s, with utilities across the country announcing accelerated retirement timelines for their remaining coal fleets. This would be an unambiguously positive story from an emissions and cost perspective if not for one complicating factor central to this book's thesis: coal plants, whatever their environmental and economic drawbacks, are dispatchable, firm resources. A coal plant with adequate fuel on site can run at full output regardless of weather, time of day, or season, for as long as its fuel supply lasts — often weeks of on-site coal inventory. This attribute has made coal retirements a persistent point of tension between environmental policy and reliability planning. NERC's reliability assessments have repeatedly flagged specific regions where the pace of coal (and other dispatchable) retirements is outrunning the pace of confirmed, in-service replacement capacity, particularly replacement capacity that can be relied upon during multi-day winter cold snaps or extended summer heat domes when both solar output and wind output can simultaneously underperform.

Several utilities and regulators have responded by delaying planned coal retirements, at least temporarily. Plants that were scheduled to close have had their retirement dates pushed back by several years in multiple states, a direct response to reliability concerns raised by grid operators and, in a growing number of cases, to requests from utilities anticipating datacenter-driven load growth that they do not yet have adequate replacement generation to serve. This is one of the more uncomfortable ironies of the current moment: the surge in electricity demand from technology companies pursuing ambitious clean energy and carbon-neutrality commitments is, in the near term, contributing to keeping some of the dirtiest power plants on the grid running longer than originally planned, simply because there is not yet enough replacement firm capacity to retire them safely.

Natural Gas: The Flexible Workhorse — and Its Fuel Supply Risk

Natural gas has become the single largest source of U.S. electricity generation, supplying roughly 40 percent of the national total. Gas plants come in two broad categories relevant to this discussion. **Combined-cycle gas plants** are highly efficient, capturing waste heat from a gas turbine to drive a secondary steam turbine, and are typically used to meet steady, ongoing "baseload" and "intermediate load" demand. **Simple-cycle gas turbines** (sometimes called "peaker plants") are less fuel-efficient but can start up and reach full output in minutes, making them ideal for meeting short, sharp demand peaks or for quickly filling in when solar or wind output drops unexpectedly.

This flexibility has made natural gas the grid's primary tool for accommodating the variability introduced by renewable energy. When a cloud bank rolls over a solar farm or wind speeds drop across a region, gas plants are frequently the resource called upon to ramp up within minutes to fill the gap. Gas has effectively become the shock absorber of the pre-battery-storage grid, a role increasingly being shared with, and in some functions displaced by, the battery storage systems discussed in Chapter 5.

But natural gas carries a reliability vulnerability that is distinct from coal's and became devastatingly clear during Winter Storm Uri in February 2021. Unlike coal plants, which typically maintain weeks of fuel inventory on site, most gas plants depend on continuous, real-time delivery of fuel through pipelines — there is generally little or no on-site fuel storage. This makes gas generation vulnerable to disruptions in the gas supply chain itself, not just to problems at the power plant. During Uri, extreme cold caused widespread freezing of wellheads, valves, and instrumentation at gas production sites across Texas — a phenomenon known as "freeze-offs" — dramatically reducing the volume of gas available to flow to power plants at precisely the moment electricity demand for heating was spiking to record levels. Compounding the problem, natural gas is also used directly for home heating, meaning gas utilities and gas-fired power plants were competing for the same shrinking supply during the crisis.

The result was catastrophic: at the height of the storm, roughly half of ERCOT's generating capacity was offline, split between frozen gas wellheads and pipelines, frozen instrumentation and wind turbines that had not been adequately winterized, and even some issues at a nuclear plant caused by cold-sensitive equipment. ERCOT was forced to implement rotating blackouts that, due to failures in the rotation process and the sheer duration of the emergency, left millions of Texans without power for hours to days in freezing conditions. More than 200 people died, many from carbon monoxide poisoning after running vehicles or generators indoors for warmth, or from hypothermia. The event triggered sweeping reforms, including new weatherization mandates for both power plants and gas infrastructure, though independent assessments in subsequent years have continued to raise questions about whether those mandates go far enough and whether compliance is adequately enforced.

Uri remains the single most important case study in modern U.S. grid reliability for a simple reason: it demonstrated, in the starkest possible terms, that "dispatchable" generation is only as reliable as its fuel supply chain, and that fuel supply chains have their own weather vulnerabilities that can align catastrophically with the exact conditions driving peak electricity demand. This lesson has become directly relevant to the AI datacenter buildout discussed later in this book, because much of the new gas-fired generation capacity being proposed to meet datacenter load growth will depend on the same pipeline infrastructure whose vulnerabilities Uri exposed.

The Existing Nuclear Fleet: A Quiet Resurgence

The United States operates roughly ninety commercial nuclear reactors, supplying about 18 to 19 percent of the nation's electricity — and, notably, closer to half of the country's carbon-free electricity generation, since nuclear plants run at very high, steady output around the clock. For much of the 2010s, the existing nuclear fleet was in slow decline: several plants closed early due to poor economics in regions with cheap natural gas and, in some deregulated markets, wholesale prices too low to cover a nuclear plant's relatively high fixed operating costs. California's Diablo Canyon and New York's Indian Point faced closure pressure driven partly by state policy preferences and public concern following the 2011 Fukushima disaster in Japan.

That trend has reversed, and the reversal is directly connected to the themes of this book. Diablo Canyon's planned closure was postponed by California state legislation specifically citing grid reliability concerns. More strikingly, several previously shuttered or scheduled-to-close nuclear plants have become the subject of restart and life-extension agreements driven substantially by AI datacenter demand — the most prominent example being the agreement between Microsoft and Constellation Energy to restart a unit at the former Three Mile Island site in Pennsylvania (rebranded the Crane Clean Energy Center) specifically to supply carbon-free power for Microsoft's datacenter operations, alongside a parallel agreement between Amazon and Talen Energy involving the Susquehanna nuclear plant. These deals, examined in more detail in Chapter 9, represent something genuinely new: technology companies directly financing the restart or life extension of existing nuclear capacity, in some cases co-locating datacenters immediately adjacent to the power plant to minimize transmission losses and interconnection queue delays.

Existing nuclear plants have also been the subject of "uprates" — engineering modifications that allow a reactor to safely generate more electricity than its original design specification, without building an entirely new plant. Cumulative uprates across the U.S. fleet have added several thousand megawatts of capacity over the past two decades, a comparatively invisible but meaningful contribution to the nation's firm generation capacity, achieved at a fraction of the cost and time required to build new nuclear plants from scratch.

Hydroelectric Power: Mature, Valuable, and Geographically Constrained

Hydroelectric power supplies roughly 6 percent of U.S. electricity, concentrated heavily in the Pacific Northwest, California, and parts of the Southeast and Northeast. Hydroelectric dams offer a combination of attributes that make them exceptionally valuable for grid reliability: they are dispatchable (output can be increased or decreased quickly by adjusting water flow through turbines), they can provide substantial frequency regulation and spinning reserve, and pumped-storage hydroelectric facilities — which pump water uphill during periods of surplus generation and release it downhill through turbines during periods of high demand — function as a form of grid-scale energy storage that predates modern batteries by decades. The United States has roughly 22 gigawatts of pumped-storage hydro capacity, some facilities dating to the mid-twentieth century, still providing some of the largest single blocks of storage capacity on the grid even as battery storage additions accelerate.

However, hydroelectric power's contribution to reliability is geographically constrained (it requires specific river and terrain conditions), environmentally contested (dam removal advocacy has grown in some regions over concerns about fish populations and river ecosystems), and increasingly subject to its own weather-related variability, as droughts in the Western United States have periodically reduced reservoir levels and hydroelectric output, most notably during severe droughts affecting the Colorado River system and California's water infrastructure in the 2020s.

Why Continued Reliance Persists

Taken together, coal, natural gas, nuclear, and hydroelectric power continued to supply roughly 60 percent of U.S. electricity generation even after two decades of rapid renewable growth, and an even higher share of the generation actually running during the system's most critical peak-demand hours. This is not evidence that the renewable transition has failed or stalled; it reflects the fact that traditional generation and renewable generation serve fundamentally different — and complementary — functions in a modern grid. Solar and wind provide low-cost energy when nature cooperates. Traditional generation provides the firm capacity, ancillary services, and around-the-clock availability that a reliable grid requires regardless of weather. Chapters 3 through 5 examine how solar, wind, and battery storage are reshaping what "firm capacity" needs to mean, and how much of it traditional plants will still need to supply as the transition continues.

The Retirement Wave in Numbers and Regional Detail

The pace of coal plant retirement varies enormously by region, and understanding that regional variation matters for grid reliability planning. The Midwest, particularly the footprint covered by MISO, has seen some of the most aggressive coal retirement activity in the country, driven by a combination of aging plants originally built in the 1960s and 1970s reaching the end of their economic lives, state clean energy mandates in states like Illinois and Michigan, and the comparative economics of replacing coal generation with wind power in a region blessed with strong wind resources, as described in Chapter 4. MISO's own reliability assessments have repeatedly flagged capacity shortfall risk in the region during multi-day winter cold events, a concern that led MISO to implement emergency cold-weather procedures and to work with member utilities on delaying certain retirements specifically to preserve reliability margin during the transition period.

The Southeast, by contrast, has seen a somewhat slower and more measured pace of coal retirement, reflecting both the vertically integrated utility structure described in this chapter's supplementary material and state regulatory commissions in states like Georgia and South Carolina that have generally supported utility plans to maintain a diversified generation portfolio, including continued coal operation for some plants well into the 2030s, specifically citing reliability considerations and the region's own now-accelerating datacenter-driven load growth described in Part III of this book.

Natural Gas Infrastructure: Pipelines as the Hidden Reliability Layer

It is worth dwelling further on the interdependence between the electricity grid and the natural gas pipeline system that Winter Storm Uri exposed so starkly, because this interdependence has become, if anything, more significant in the years since Uri as natural gas generation has continued to expand its share of the national generation mix. Natural gas moves through an extensive network of interstate and intrastate pipelines, compressor stations (which use their own natural gas-fired or electric-powered compressors to keep gas moving through the pipeline at adequate pressure), and underground storage facilities that can hold gas in reserve during periods of peak demand.

This system was originally designed and built primarily to serve natural gas utilities providing direct heating service to homes and businesses, a demand pattern that peaks predictably during winter cold snaps. The rapid growth of gas-fired electricity generation, particularly the growth of gas plants specifically built to provide fast-ramping support for renewable energy integration as described in Chapter 3, has added an additional, and in some respects less predictable, source of demand on the same pipeline infrastructure, one that can spike suddenly when renewable output unexpectedly drops rather than following the more predictable seasonal pattern of heating demand. During Winter Storm Uri, this compounding effect was severe: extreme cold simultaneously froze wellhead and pipeline equipment (reducing gas supply), spiked heating demand from gas utility customers (increasing demand for the same shrinking supply), and spiked electricity demand for heating that in turn increased demand for gas-fired power generation, creating a demand spike from two directions competing for a supply that was simultaneously shrinking due to the same cold weather.

Post-Uri reforms have attempted to address this vulnerability through several mechanisms: mandatory weatherization standards for both gas production and gas-fired power plant equipment, designation of certain natural gas infrastructure as "critical" during energy emergencies (intended to prioritize its electricity supply, preventing the circular failure mode in which power outages at compressor stations further reduced gas flow, worsening the very power shortage the gas was needed to address), and improved coordination and information-sharing protocols between electric grid operators and natural gas pipeline operators, who had historically operated as largely separate industries with limited real-time coordination despite their deep operational interdependence. Independent assessments conducted in the years following Uri have generally found meaningful improvement in weatherization compliance, while also noting continued gaps in enforcement consistency and in the pace of compressor station and other critical gas infrastructure electrification or backup power hardening.

Nuclear Economics in Deregulated Markets

The nuclear fleet's economic challenges in deregulated ISO/RTO markets, mentioned briefly in this chapter, deserve fuller explanation because they directly set up the datacenter-nuclear power purchase agreements discussed in Chapter 9. In a deregulated wholesale market, a nuclear plant earns revenue by selling its electricity output at the prevailing wholesale market price for each hour, a price set by the marginal cost of the most expensive generator needed to meet demand in that hour — frequently a natural gas plant. Nuclear plants have very high fixed costs (the capital cost of construction, spread over the plant's operating life, along with substantial fixed staffing, security, and regulatory compliance costs) but relatively low marginal costs (the cost of the uranium fuel itself is a small fraction of a nuclear plant's total costs). This cost structure means nuclear plants are economically vulnerable in markets where wholesale prices are frequently low — a situation that became increasingly common through the 2010s as cheap natural gas and, later, zero-marginal-cost wind and solar generation pushed down average wholesale prices in many regions, even though nuclear plants still needed to recover their substantial fixed costs.

Several states responded to this vulnerability by creating "zero-emission credit" programs, state-level subsidy mechanisms that pay existing nuclear plants for their carbon-free attributes, functioning similarly to the renewable energy credits states have long used to support wind and solar development, and explicitly designed to prevent premature nuclear plant closures purely for market-driven economic reasons, in states including New York, Illinois, New Jersey, and Connecticut. These state-level interventions kept several nuclear plants operating through the 2010s and early 2020s that might otherwise have closed, setting the stage for the datacenter-driven demand for nuclear power purchase agreements that emerged later in the decade — agreements that, unlike the zero-emission credit programs, involve direct, long-term contracts between individual corporate customers and nuclear plant owners, a genuinely new financing model for existing nuclear capacity that Chapter 9 examines in detail.

The Human and Economic Dimension of Coal Retirement

Beyond the reliability engineering questions this chapter has emphasized, coal plant retirements carry a significant human and economic dimension worth acknowledging. Coal plants, and the coal mines that supply them, have historically anchored the economies of numerous rural communities, particularly in Appalachia, the Illinois Basin, and Wyoming's Powder River Basin, providing well-compensated jobs and substantial local tax revenue that fund schools, infrastructure, and local government services in communities that often have limited alternative economic anchors. Plant and mine closures have accordingly triggered significant local economic disruption in many of these communities, a dimension of the energy transition that has become an increasingly prominent focus of federal and state policy attention, including targeted federal funding programs aimed at supporting economic diversification and workforce transition in coal-dependent communities, and the kind of direct site-reuse strategies exemplified by TerraPower's decision to build its Wyoming small modular reactor demonstration project at a retiring coal plant site specifically to preserve local jobs and tax base, discussed further in Chapter 9.

This human dimension adds an important layer of political complexity to the reliability-driven retirement delays discussed in this chapter's main text: in some cases, the same communities and political constituencies advocating for delayed coal retirement on reliability grounds are also those most directly affected by the broader economic transition away from coal, creating a alignment of interests between reliability-focused utility planners and economically motivated local and state political leaders that has, in practice, reinforced the trend toward delayed rather than accelerated coal retirement in several states over the mid-2020s.

Combined Heat and Power and Industrial Cogeneration

A traditional generation category deserving brief mention, distinct from the utility-scale power plants that dominate this chapter's discussion, is **combined heat and power (CHP)**, also called cogeneration – industrial and commercial facilities that generate their own electricity on site, typically using natural gas, while also capturing the waste heat from that generation process for use in industrial processes or building heating, achieving considerably higher overall energy efficiency than generating electricity and heat separately. CHP has long been used by large industrial facilities, universities, hospitals, and other campuses with substantial simultaneous electricity and heat demand, and it represents an early, if less discussed, precedent for the kind of behind-the-meter generation strategies datacenter developers have increasingly pursued, as described in Chapter 7. Some datacenter operators have explored CHP-style configurations that capture waste heat from on-site generation for use in facility heating or even, in some experimental projects, for district heating supplied to nearby communities, illustrating a point of continuity between this older industrial generation model and the newer datacenter-specific generation strategies discussed later in this book.

Fuel Diversity as Explicit Reliability Policy

Several of the reliability concerns discussed throughout this chapter have led utilities and regulators to increasingly treat "fuel diversity" – maintaining a generation portfolio that draws on multiple distinct fuel sources and supply chains, rather than concentrating too heavily on any single fuel type – as an explicit reliability policy goal in its own right, separate from and complementary to the reserve margin and capacity adequacy concepts introduced in

Chapter 1. The logic follows directly from the Winter Storm Uri experience described in this chapter: because natural gas, coal, and even wind and solar each have distinct weather-related vulnerabilities that can align differently with different types of extreme weather events, a generation portfolio drawing on a genuinely diverse mix of fuel types and technologies is less likely to experience a correlated, simultaneous failure across its entire capacity than a portfolio concentrated in any single fuel source, even if that single fuel source is, on average, the cheapest or most reliable option under normal conditions. This fuel diversity principle has become an explicit factor in several utilities' integrated resource planning processes and in some state regulatory guidance, representing a documented lesson learned directly from the Uri crisis and reinforcing this chapter's broader argument that continued reliance on a diversified traditional generation base, even amid the broader renewable transition, remains a deliberate and defensible reliability strategy rather than simply an artifact of slow decarbonization progress.

Emergency Fuel Reserves and Strategic Stockpiling

A final policy tool worth noting in the traditional generation discussion of this chapter involves emergency fuel reserve requirements, an idea that gained renewed attention following Winter Storm Uri's demonstration of natural gas supply chain fragility. Some states and regional reliability organizations have explored requiring power plants, particularly those serving critical winter reliability functions, to maintain a minimum on-site fuel inventory or firm fuel supply contract, analogous to the multi-week coal stockpiles many coal plants have traditionally maintained on site, discussed earlier in this chapter, but applied to natural gas generators that have historically operated with little or no on-site fuel storage of their own. Implementing such requirements for gas plants is considerably more complex than for coal, since natural gas cannot be as easily or cheaply stockpiled on site in large quantities without specialized (and expensive) liquefied natural gas storage infrastructure, meaning most proposals in this area have instead focused on firm, priority-delivery fuel supply contracts and improved coordination protocols between gas pipeline operators and electricity generators, rather than true physical on-site stockpiling. These measures remain a work in progress across different states and regions as of the mid-2020s, illustrating that the fuel supply chain vulnerability Uri exposed, discussed at length earlier in this chapter, has not been fully resolved even several years after the crisis that revealed it, and remains an area of continued regulatory attention as the broader reliability challenges described throughout this book continue to unfold.

PART II

THE RENEWABLE TRANSITION AND ITS RELIABILITY TRADE-OFFS

Chapter 3:

Solar's Rapid Rise

Utility-scale solar power has grown faster than almost any energy technology in American history. In the early 2010s, utility-scale solar was a marginal contributor to the national generation mix, and most Americans who thought about solar power at all pictured rooftop panels on a neighbor's house. By the mid-2020s, utility-scale solar farms — sprawling installations covering hundreds or even thousands of acres with photovoltaic panels — had become one of the largest sources of new generating capacity added to the grid each year, exceeding new natural gas capacity additions in most years since the early 2020s. This chapter examines why solar grew so quickly, the operational challenges its growth has created, and how those challenges are reshaping grid operations.

The Cost Collapse That Made It Possible

The single most important fact about solar power's rise is the extraordinary decline in its cost. The price of photovoltaic solar modules fell by roughly 90 percent between 2010 and the early 2020s, driven by manufacturing scale economies (heavily concentrated in Chinese manufacturing, a connection explored further in Part V of this book), improvements in panel efficiency, and falling costs for the balance-of-system components — inverters, racking, wiring, and installation labor — that make up a utility-scale solar project. By the early 2020s, unsubsidized utility-scale solar had become the cheapest source of new electricity generation in much of the United States on a pure levelized-cost basis, even before accounting for federal tax credits that further improved its economics.

This cost collapse was reinforced by federal policy, particularly the investment tax credit for solar projects, which has existed in various forms since 2006 and was extended and expanded by the Inflation Reduction Act of 2022. State-level renewable portfolio standards, requiring utilities to source a growing percentage of their electricity from renewable sources, provided an additional durable demand signal that gave developers confidence to build large projects. The combination of falling costs and durable policy support produced a self-reinforcing cycle: falling costs drove more deployment, more deployment drove further manufacturing scale and technology refinement, and further refinement drove costs down again.

The Interconnection Queue Bottleneck

Solar's growth has not been limited by technology cost or even by available land in most regions; it has been limited by a much less visible bottleneck: the process by which new generation projects connect to the grid. Before any new power plant — solar, wind, gas, storage, or otherwise — can begin operating, it must undergo an interconnection study process, in which the relevant transmission owner or ISO analyzes whether the project can be safely connected without causing reliability problems on the existing grid, and what upgrades (new transmission lines, substation equipment, or protective relaying) might be required to accommodate it.

This process has become severely backlogged across nearly every region of the country. By the mid-2020s, the total capacity of projects sitting in U.S. interconnection queues — the vast majority of it solar, wind, and battery storage — had grown to several times the total installed generating capacity of the entire country, with average time from initial interconnection request to commercial operation stretching to five years or longer in many regions, up from roughly two years in the early 2000s. Much of this delay stems from a "first-come, first-served" queue design in which each new project's interconnection study must account for every project that requested interconnection before it, creating cascading re-studies whenever an earlier project in the queue changes its specifications or withdraws.

Federal regulators, through the Federal Energy Regulatory Commission, issued major queue reform rules in 2023 aimed at moving toward more of a "first-ready, first-served" cluster study approach, and individual ISOs including PJM, MISO, and CAISO have implemented reforms of their own. These reforms have begun to reduce backlogs in some regions, but the queue remains one of the most significant practical constraints on how quickly new solar (and other) generation can actually reach the grid — a constraint that matters enormously for the demand-growth story told in Part III, since utilities racing to serve new datacenter load cannot simply wave a wand and add gigawatts of new solar capacity; they must navigate a multi-year interconnection process even after a project is otherwise ready to build.

The Duck Curve Problem

Solar power's most famous operational challenge has a memorable name: the **duck curve**, a term coined by CAISO analysts around 2013 to describe the shape of a graph showing net electricity demand (total demand minus renewable generation) across a typical day in a grid with high solar penetration. In the late morning and early afternoon, as solar output ramps up, net demand — the amount of electricity that traditional generation must supply — drops sharply, forming the flat "belly" of the duck. But as the sun sets in the late afternoon and early evening, solar output collapses precisely as residential electricity demand is rising (people returning home, turning on lights, air conditioning, and appliances), creating an extremely steep ramp in the amount of power that other generation sources must supply within a compressed few hours — the duck's steep "neck."

This evening ramp has become one of the defining operational challenges for grid operators in high-solar regions like California. In some years, CAISO has needed to bring online several thousand additional megawatts of generation within a three-to-four-hour window each evening, a ramping requirement that traditional baseload plants (which are optimized to run at steady output, not to cycle up and down quickly) handle poorly, and that has historically been met primarily by natural gas peaker plants. As Chapter 5 will discuss in detail, battery storage has emerged as an unusually well-suited tool for exactly this problem: batteries can charge during the solar-abundant midday hours, when wholesale electricity prices are often driven down toward zero or even negative by an oversupply of solar power, and then discharge during the evening ramp, effectively "moving" midday solar generation into the evening hours when it is actually needed.

Curtailment: When Solar Power Goes to Waste

A less visible but increasingly significant consequence of solar's rapid growth is **curtailment** – the deliberate reduction of solar (or wind) output below what the resource could otherwise generate, because there is no demand for the power, no transmission capacity available to move it elsewhere, or because system reliability requires reducing output from renewable resources to make room for the minimum output of other generators that must stay online for stability reasons.

Curtailment has grown substantially in high-solar-penetration regions. California curtailed several million megawatt-hours of solar and wind generation in recent years – enough electricity to power hundreds of thousands of homes for a year – with curtailment concentrated overwhelmingly in the spring months, when electricity demand is relatively low (mild weather reduces both heating and cooling loads) but solar output remains high. Texas has experienced a similar and rapidly growing curtailment problem as its own utility-scale solar fleet has expanded dramatically, compounded by transmission constraints that prevent power generated in West Texas, where much of the state's wind and solar capacity is concentrated, from reaching demand centers around Dallas-Fort Worth, Houston, and Austin.

Curtailment represents a genuine economic inefficiency — capital-intensive generating capacity sitting idle when it could be producing value — and it has become one of the strongest arguments for the batteries described in Chapter 5, for expanded transmission capacity, and, more recently, for deliberately siting new large flexible loads (including datacenters) in locations and operating patterns that can absorb otherwise-curtailed renewable power. Some technology companies have begun exploring datacenter operating strategies that flex their computing workloads to periods of abundant, cheap renewable generation, though this "load-following" approach remains limited in practice given the largely continuous, latency-sensitive nature of most current AI training and inference workloads, a tension explored further in Chapter 7.

Solar's Reliability Contribution: Real but Limited

None of these challenges mean solar power lacks reliability value — it simply means that value must be measured carefully and honestly. Grid planners assess a resource's contribution to reliability using a metric called the **effective load-carrying capability (ELCC)**, which estimates how much a given resource actually reduces the probability of a supply shortfall during the hours when the system is most stressed, rather than simply counting its full nameplate capacity. Solar's ELCC is generally strong during summer afternoon peak-demand hours in most U.S. regions, since air conditioning demand and solar output are correlated (both driven by sunny, hot weather). But solar's ELCC value declines as more solar is added to a system, a phenomenon sometimes called "value deflation" — the first gigawatt of solar capacity added to a grid is more valuable, from a reliability standpoint, than the tenth or the twentieth gigawatt, because additional solar capacity increasingly overlaps with hours when the system already has abundant solar output and does not need more. Perhaps most importantly, solar's ELCC value drops sharply for winter peak demand and for early-evening summer peak hours after sunset — precisely the hours when the duck curve's steep ramp occurs and when several major grid emergencies, including parts of Winter Storm Uri, have occurred.

This is why grid planners increasingly describe solar as an "energy resource" rather than fully a "capacity resource": tremendously valuable for reducing fuel costs and emissions across the year, but only partially substitutable for the firm, dispatchable capacity described in Chapter 2, and requiring complementary resources — natural gas, batteries, and potentially the emerging nuclear technologies described in Chapter 9 — to fill the gaps that solar's daily and seasonal patterns leave behind.

Community Solar and Distributed Generation

While this chapter has focused primarily on utility-scale solar — large, centrally sited projects that interconnect directly to the transmission or high-voltage distribution system — it is worth noting the continued growth of smaller-scale distributed solar, including rooftop residential and commercial solar installations and "community solar" projects that allow customers without suitable rooftops of their own to subscribe to a share of a nearby shared solar installation's output. Distributed solar has grown steadily alongside utility-scale solar, though at a more modest pace, and it introduces its own distinct set of grid integration challenges, primarily at the distribution system level rather than the transmission and bulk-power system level that dominates this book's focus.

As distributed solar penetration has grown in certain neighborhoods and circuits, utilities have had to grapple with reverse power flow (electricity flowing from a residential circuit back toward the substation, rather than the traditional one-directional flow described in Chapter 1), voltage regulation challenges on circuits with high concentrations of solar installations, and the broader question of how to fairly compensate distributed solar owners for the electricity they contribute to the grid, a subject of extensive and often contentious regulatory debate over "net metering" policy in numerous states. These distribution-level challenges, while significant for utility distribution planning, are generally distinct from and smaller in overall reliability significance than the bulk-system, utility-scale solar integration challenges — the duck curve and curtailment described in the main body of this chapter — that matter most for the system-wide reliability questions this book addresses.

Agrivoltaics and Land Use Considerations

The rapid pace of utility-scale solar deployment has increasingly run into land-use tensions, particularly in agricultural regions where large solar projects compete with farmland for the same relatively flat, well-drained, grid-accessible land that both uses favor. This tension has spurred growing interest in **agrivoltaics** — the practice of co-locating solar panels with continued agricultural use of the same land, either through elevated panel mounting that allows crops or grazing animals to continue beneath the panels, or through careful project design that dedicates only a portion of a given parcel to solar while preserving the remainder for continued farming. Agrivoltaic approaches remain a smaller share of total utility-scale solar deployment than conventional ground-mounted projects, but they have grown in prominence as a response to local community and agricultural industry concerns about solar development's cumulative land-use footprint, and several states have begun incorporating agrivoltaic considerations into their solar project permitting and siting review processes.

Beyond agrivoltaics, utility-scale solar developers have increasingly prioritized siting projects on disturbed or otherwise lower-conflict land — former industrial sites, closed landfills, and, notably, the sites of retiring coal plants, which often already have valuable existing transmission interconnection infrastructure that a new solar project can reuse, avoiding some of the interconnection queue delays described in this chapter's main discussion. This pattern of reusing retiring fossil generation sites for new renewable or storage projects — sometimes informally called "brownfield-to-green" redevelopment — has become an increasingly common and cost-effective siting strategy, and it parallels the coal-to-nuclear site reuse strategy TerraPower has pursued for its Wyoming small modular reactor project, discussed in Chapter 9.

The Growing Role of Solar-Plus-Storage Hybrid Projects

An increasingly dominant trend in utility-scale solar development, closely connected to the battery storage discussion of Chapter 5, is the **solar-plus-storage hybrid project**, in which a solar array and a battery storage system are developed, interconnected, and often operated together at a single site, sharing a single interconnection point and, frequently, a single set of power electronics and grid interconnection equipment. This co-location approach offers several advantages over developing solar and storage as entirely separate projects: it can reduce total interconnection costs and timeline by sharing a single interconnection study and approval process, it allows the battery to charge directly from the co-located solar array without incurring additional transmission or distribution charges that would apply if the battery instead charged from the broader grid, and it allows project developers and their financing partners to optimize the combined solar-and-storage system's operation as a single integrated resource rather than two separately optimized assets.

The share of new utility-scale solar capacity being developed as part of a hybrid solar-plus-storage project has grown substantially through the mid-2020s, particularly in California and other high-solar-penetration markets where the storage component's value in addressing the duck curve's evening ramp, described earlier in this chapter, is most pronounced. This trend represents one of the clearest pieces of evidence that the solar and storage stories told in this chapter and the next are not really two separate stories at all, but increasingly a single, integrated development pattern in which storage has become close to a standard, expected complement to new utility-scale solar development, rather than a separate technology pursued independently.

Solar Forecasting: The Unsung Enabler of High Penetration

Much as improved wind forecasting has enabled greater wind penetration, as discussed in Chapter 4, improved solar forecasting has played an underappreciated but essential role in enabling the high levels of solar penetration described throughout this chapter. Grid operators including CAISO rely on sophisticated satellite-imagery-based and ground-sensor-based cloud-tracking systems, combined with numerical weather prediction models, to forecast solar output at increasingly fine geographic and temporal resolution, allowing operators to anticipate the duck curve's evening ramp and shorter-term cloud-driven output fluctuations with enough lead time to schedule complementary generation and, increasingly, battery dispatch accordingly. Forecast accuracy has improved substantially over the past decade, though solar forecasting remains inherently more challenging on short (minutes-to-hours) timescales than on longer (day-ahead) timescales, since cloud formation and movement remain among the more difficult phenomena to predict precisely, particularly in regions with complex terrain or coastal marine-layer cloud patterns that can behave less predictably than clear, cloudless desert conditions.

Solar Manufacturing Policy and the Domestic Content Debate

The Inflation Reduction Act's clean energy tax credits, discussed earlier in this chapter, include specific bonus credit provisions for projects using domestically manufactured solar components, part of a broader federal policy effort to rebuild U.S. solar manufacturing capacity that had largely migrated overseas, principally to China, over the preceding two decades as Chinese manufacturers achieved the scale and cost advantages described in Chapter 11. This domestic content policy has spurred a meaningful expansion of U.S. solar panel and component manufacturing capacity, including new or expanded manufacturing facilities in Georgia, Ohio, and several other states, though U.S. manufacturers continue to face significant cost competition from established Chinese manufacturing scale, and the domestic solar manufacturing supply chain remains incomplete in certain upstream stages, particularly polysilicon refining and certain specialized manufacturing equipment, where Chinese and other Asian manufacturers continue to hold a dominant global market position that a few years of U.S. policy incentive have not yet fully overcome. This tension between the goal of rapid, low-cost solar deployment (best served, in the near term, by continued access to lower-cost imported components) and the goal of rebuilding domestic manufacturing capacity and supply chain resilience (which may involve somewhat higher near-term costs) represents a genuine policy trade-off that has shaped the specific design of federal solar incentive programs and that connects directly to the broader U.S.-China supply chain dependency discussion in Part V of this book.

Floating Solar: An Emerging Niche Application

A smaller but growing niche within utility-scale solar development involves **floating solar** installations, in which solar panels are mounted on floating platforms deployed on reservoirs, retention ponds, or other bodies of water rather than on land. Floating solar offers several potential advantages: it avoids the land-use tensions with agriculture discussed in this chapter's earlier supplementary material, the water beneath the panels can help keep panel temperatures lower (improving electricity generation efficiency, since photovoltaic panels lose some efficiency as their operating temperature rises), and in some configurations, the shading provided by the panels can reduce evaporative water loss from the reservoir, an increasingly valued benefit in drought-prone regions of the Western United States. Floating solar remains a considerably smaller category of overall utility-scale solar deployment than conventional ground-mounted installations, constrained by higher installation costs and the more limited number of suitable water bodies, particularly those owned by utilities or public agencies willing to lease their surface for this purpose, but it has grown steadily as a specialized complement to the broader utility-scale solar buildout described throughout this chapter, particularly at sites including hydroelectric reservoirs, where floating solar can share existing transmission interconnection infrastructure with the underlying hydroelectric facility, echoing the interconnection-reuse advantage discussed for repowered wind farms in Chapter 4 and coal-to-nuclear site conversions in Chapter 9.

Solar's Contribution to Emergency Preparedness at the Residential Level

Beyond the utility-scale focus of this chapter, it is worth noting that residential solar paired with home battery backup, mentioned briefly in Chapter 13's discussion of individual household resilience, has grown into a meaningful personal reliability strategy for homeowners in regions facing elevated outage risk, whether from the wildfire-related public safety power shutoffs that have affected parts of California or from the extreme weather events discussed throughout this book. While residential solar-plus-storage systems remain a comparatively small contributor to bulk system reliability relative to the utility-scale resources this book has emphasized, their growing adoption reflects a broader public response to the same underlying reliability concerns motivating the utility-scale strategies described throughout this chapter and the next, and represents a visible, tangible way many households have chosen to hedge against the very risks this book examines at the system level.

Chapter 4:

Wind Power's Regional Story

If solar power's story is one of a technology that grew explosively once its costs collapsed, wind power's story is one of a technology that matured a decade earlier and now occupies a large, stable, but geographically distinctive role in the American generation mix. Wind power supplies roughly 10 percent of U.S. electricity generation nationally, but that national figure masks enormous regional variation: in parts of the wind-rich Great Plains, wind power supplies well over a third of annual electricity generation, while in much of the Southeast it supplies essentially none.

The Great Plains Wind Corridor

The single most important geographic fact about American wind power is the existence of an enormous, high-quality wind resource running down the center of the country, roughly from North Dakota through South Dakota, Nebraska, Kansas, Oklahoma, and into the Texas Panhandle and West Texas. This corridor benefits from consistent, strong surface winds created by the interaction of the Rocky Mountains, the flat, unobstructed terrain of the Great Plains, and continental-scale temperature and pressure patterns. Texas alone has more installed wind capacity than all but a handful of entire countries, concentrated heavily in West Texas and the Texas Panhandle. Iowa generates more than half of its annual electricity from wind power, one of the highest wind-penetration rates of any U.S. state.

This concentration of wind resources far from the major population centers that consume the electricity — Dallas-Fort Worth, Houston, San Antonio, and Austin for Texas wind; Chicago and other Midwest population centers for Great Plains wind more broadly — has created a persistent and expensive transmission challenge. Texas addressed this challenge through a landmark infrastructure program called the Competitive Renewable Energy Zones (CREZ) initiative, approved in 2005, which built roughly 3,600 miles of new transmission lines specifically to move wind power from West Texas to the state's population centers, at a cost of several billion dollars, completed around 2013. CREZ is often cited as a model for proactive, planned transmission investment specifically enabling renewable energy development, though similar large-scale transmission buildouts have proven far more difficult to execute in other parts of the country, where transmission projects must cross multiple state jurisdictions, each with its own permitting authority and often its own political resistance to hosting transmission infrastructure for power that will primarily benefit a different state.

Offshore Wind: The Northeast's Emerging (and Struggling) Resource

While onshore wind has flourished in the center of the country, the Northeast and Mid-Atlantic states — dense population centers with limited land available for large onshore wind or solar projects, and states with aggressive clean energy mandates — have looked instead to offshore wind, harnessing the strong, consistent winds available over the Atlantic Ocean. States including Massachusetts, Rhode Island, New York, New Jersey, and Virginia have pursued offshore wind development aggressively through state procurement mandates and dedicated offshore wind targets, with several projects reaching commercial operation in the early-to-mid 2020s, including Vineyard Wind off Massachusetts and the Coastal Virginia Offshore Wind project, developed by Dominion Energy, which stands as the largest offshore wind project under development in U.S. waters.

Offshore wind has faced a considerably rockier path than onshore wind or solar, however. Rising interest rates, supply chain inflation, and turbine manufacturing challenges caused several major offshore wind projects to be canceled, renegotiated, or delayed in the early-to-mid

2020s, with some developers taking substantial financial write-downs and, in a few cases, walking away from previously signed power purchase agreements rather than complete projects at costs that had risen well above their original contracted price. Federal permitting and leasing policy for offshore wind has also shifted significantly with changes in presidential administrations, creating a degree of policy uncertainty that has made project financing more difficult. As of the mid-2020s, offshore wind remains a real and growing, but considerably smaller and more uncertain, contributor to the national generation mix than onshore wind, with its future trajectory more dependent on federal and state policy continuity than any other major renewable technology discussed in this book.

Wind's Intermittency Challenge — Different in Character from Solar's

Wind power is intermittent in a different way than solar power. Solar's variability is highly predictable on a daily basis (the sun reliably rises and sets at known times) but variable on shorter timescales due to cloud cover, and essentially zero at night. Wind's variability is less tied to a predictable daily cycle and more driven by weather systems that can persist for hours to days, meaning wind output can remain low across an entire region for extended periods — a phenomenon sometimes informally called a "wind drought" — precisely during large-scale weather events such as summer heat domes (often associated with stagnant high-pressure systems that produce both extreme heat and low wind) or, more alarmingly for reliability planners, winter cold snaps in which extreme cold can be accompanied by icing on turbine blades that forces them offline for safety reasons, exactly as electricity demand for heating spikes. This pattern was a significant contributor to the generation shortfall during Winter Storm Uri, when a substantial share of ERCOT's wind fleet was unable to generate due to iced turbines and, to a lesser degree, low wind speeds, at the same moment gas and coal generation was also failing.

Forecasting technology for wind power has improved substantially over the past decade, with grid operators now using sophisticated numerical weather prediction models combined with real-time turbine-level data to forecast wind output hours to days in advance with meaningfully better accuracy than in the 2000s and early 2010s. Improved forecasting does not eliminate wind's variability, but it allows grid operators to schedule complementary generation resources — and increasingly, battery storage — more efficiently in anticipation of expected wind output changes, reducing (though not eliminating) the amount of costly, fast-reacting reserve capacity that must be held in reserve purely to cover forecast errors.

Wind-Solar Complementarity

One of the more encouraging operational findings from a decade of high renewable penetration is that wind and solar generation patterns are often meaningfully complementary across both daily and seasonal timescales. In many regions, wind output tends to be strongest overnight and in the shoulder seasons (spring and fall), while solar output is naturally concentrated in daytime hours and is typically strongest in summer. This complementarity means that a generation portfolio combining both wind and solar tends to produce a smoother, more consistent combined output profile than either resource alone, reducing (though again, not eliminating) the magnitude of the ramping and reserve challenges described for solar alone in Chapter 3.

This complementarity has become an explicit factor in resource planning. Utilities and ISOs increasingly model prospective new generation portfolios using hourly, chronological simulations that capture how wind, solar, and other resources interact across an entire year, rather than relying on older planning approaches that considered only a handful of peak-demand snapshot hours. These more sophisticated modeling approaches have generally reinforced the conclusion that diversified renewable portfolios, combined with storage and firm backup capacity, can achieve high renewable penetration levels while maintaining reliability — but they have also reinforced the conclusion that a renewable-heavy grid cannot rely on wind and solar alone, and that the firm capacity and storage resources described in Chapters 2 and 5 remain essential complements, not optional extras, to the wind and solar buildout described in this chapter and the last.

Transmission as the Binding Constraint

If there is a single unifying lesson from both the solar and wind chapters, it is that the American grid's greatest renewable energy constraint by the mid-2020s was not technology cost, resource availability, or even public acceptance — it was transmission capacity, both within regions (moving power from remote wind and solar zones to demand centers) and between regions (allowing one part of the country experiencing a renewable energy surplus to export it to a neighboring region experiencing a shortfall, or vice versa). Studies conducted by national laboratories and independent grid researchers have repeatedly found that expanded interregional transmission — particularly connections between the Eastern and Western Interconnections and between wind-rich and demand-dense regions — would provide substantial reliability and cost benefits by allowing renewable generation and, importantly, reserve capacity to be shared across much larger geographic footprints, smoothing out the local weather variability that most challenges a single region operating in relative isolation.

Yet interregional transmission projects in the United States routinely take a decade or more to permit and build, hampered by multi-state jurisdictional approval processes, local opposition to new transmission corridors, and a cost-allocation problem in which the states or utilities that would host new transmission lines do not always capture a proportional share of the benefits, reducing their political incentive to approve projects primarily benefiting other regions. This transmission bottleneck, introduced here, becomes directly relevant again in Chapter 7, where it emerges as one of the central obstacles utilities face in trying to rapidly expand generation and delivery capacity to meet AI datacenter demand growth, and again in Chapter 12, where it forms part of the comparison between the pace of U.S. and Chinese grid infrastructure development.

Repowering: Extracting More from Existing Wind Farms

As the oldest utility-scale wind farms in the United States, some dating to the late 1990s and early 2000s, have approached the end of their original design lives, the wind industry has increasingly turned to **repowering** — replacing older turbines at an existing wind farm site with newer, larger, more efficient turbine models, while reusing the site's existing land lease agreements, access roads, and, critically, its existing transmission interconnection rights. Because modern wind turbines are substantially larger and more efficient than the turbines installed even a decade earlier — taller towers reach into stronger and more consistent wind resources higher above the ground, and longer blades sweep a larger area to capture more energy from a given wind speed — repowering an existing site can significantly increase its total energy output without requiring the multi-year interconnection queue process, described in Chapter 3's main discussion of solar (and applicable equally to wind), that a genuinely new project at a new site would require.

Repowering has become an increasingly significant source of effective new wind generation capacity in mature wind markets like Iowa, Texas, and California, and it illustrates a broader pattern relevant throughout this book: given the substantial interconnection queue and permitting delays affecting genuinely new generation projects of every technology type, strategies that extract more value from existing interconnected infrastructure — repowering wind farms, reconductoring transmission lines as described in

Chapter 7, uprating existing nuclear plants as described in Chapter 2 — have become increasingly attractive relative to entirely new projects requiring a fresh interconnection process, precisely because they can often be completed on a considerably faster timeline.

Community and Environmental Opposition to Wind Development

Despite wind power's strong underlying economics in much of the country, wind development has faced persistent and, in some regions, intensifying local opposition, for reasons that mix aesthetic concerns (opposition to the visual impact of large turbines on rural landscapes), noise and shadow-flicker concerns among nearby residents, wildlife impact concerns (particularly regarding bird and bat mortality, and, for offshore projects, concerns about impacts on marine mammals including the endangered North Atlantic right whale), and, in some cases, broader political polarization around renewable energy and climate policy that has made wind project permitting a more contested and politically charged process in some states and counties than it once was.

These tensions have contributed to a growing number of county and township-level moratoria or outright bans on new wind development in parts of the Midwest and other regions, even in states with generally favorable wind resources and historically supportive state-level renewable energy policy. This local, county-level opposition represents a distinct and, in some respects, more granular obstacle than the state and federal permitting challenges described elsewhere in this book, since it can effectively foreclose wind development in specific localities regardless of the broader state or national policy environment, and it has pushed some wind developers to focus increasingly on offshore wind (despite that technology's own distinct challenges, described in this chapter's main discussion) or on repowering existing sites, described above, which face less new-siting-related local opposition than genuinely new onshore wind farm locations.

Wind Power's Contribution to Winter Reliability: A More Complicated Picture Than Often Assumed

Public discussion of wind power's reliability contribution during winter events has sometimes been oversimplified, in ways this book aims to correct. During Winter Storm Uri, wind turbine icing and low wind speeds did meaningfully reduce ERCOT's wind generation output relative to what would have been expected under normal conditions, a fact seized upon by some commentators to argue that wind power is fundamentally unreliable. But a fuller accounting of the event shows that wind's underperformance was a proportionally smaller contributor to the overall generation shortfall than the failures affecting natural gas and, to a lesser extent, coal generation, which represented a much larger share of ERCOT's total generation capacity and experienced comparably severe, and in some respects more severe, proportional output reductions due to the same extreme cold, chiefly through the frozen gas supply chain issues described in Chapter 2's supplementary discussion.

Subsequent winter events have shown a more nuanced pattern: wind output during winter cold snaps varies considerably depending on the specific weather system involved, performing reasonably well during some events (when strong winds accompany cold weather, a not-uncommon combination given the same large-scale weather systems that bring cold air often also bring strong winds) and poorly during others (when cold weather is associated with stagnant high-pressure systems producing minimal wind, or when ice accumulation specifically affects turbine blades). This variability underscores a theme repeated throughout this book: no single generation technology, including natural gas, is reliably available under all extreme weather conditions, and genuine reliability requires a diversified portfolio of resources along with adequate reserve margins sized to account for the possibility that several resource types may underperform simultaneously during the most severe events — precisely the kind of correlated, multi-resource-type failure that occurred during Uri and that reliability planners now explicitly model when calculating appropriate reserve margins for extreme winter conditions.

Wind Power Purchase Agreements and Corporate Procurement

A significant share of U.S. wind development over the past decade has been financed not primarily through utility-owned generation or state renewable portfolio standard compliance, but through corporate power purchase agreements, in which large corporations — including many of the same technology companies whose datacenter electricity demand is described throughout Part III of this book — contract directly with wind (and solar) project developers to purchase the output of a specific project over a long-term contract, often fifteen to twenty years, providing the developer with the revenue certainty needed to secure project financing. Technology companies including Google, Meta, and Amazon have each been among the largest corporate purchasers of wind and solar power in the country for much of the past decade, motivated by corporate sustainability commitments that, in many cases, predate and are analytically distinct from the more recent, datacenter-electricity-specific nuclear power purchase agreements discussed in Chapter 9, though both reflect the same broader corporate interest in securing long-term, cost-predictable, and increasingly carbon-free electricity supply.

This corporate procurement model has grown into one of the most significant financing mechanisms for new wind (and solar) development in the country, in some years accounting for a share of new capacity additions comparable to or exceeding that driven directly by state renewable portfolio standard compliance. It is worth noting an important nuance in how these corporate agreements typically function: because the purchasing corporation and the wind or solar project are often not physically connected in a direct, "behind the meter" sense, and because the corporation's actual electricity consumption (for a datacenter or other facility) typically draws from the local grid rather than directly from the contracted wind or solar project, these arrangements are usually structured as "virtual" power purchase agreements, in which the corporation contracts for the project's electricity output on a financial basis, receiving payment when wholesale market prices are below the contracted price and making payment when prices are above it, while the physical electricity generated by the project is sold into the wholesale market and the corporation's actual facility draws power from the grid in the ordinary way described in Chapter 1. This financial structure allows corporations to support new renewable development and make credible carbon-reduction claims without requiring the kind of direct physical interconnection that the co-located nuclear and behind-the-meter generation arrangements discussed in Chapters 7 and 9 involve, illustrating two genuinely distinct models — financial contracting versus direct physical co-location — that technology companies have pursued in parallel to secure the electricity supply their operations and sustainability commitments require.

The Meteorological Science Behind Wind Resource Assessment

Before any wind farm is built, developers conduct extensive wind resource assessment, typically involving one to several years of on-site meteorological data collection using tall measurement towers or, increasingly, remote-sensing technologies including LiDAR (light detection and ranging) systems that can measure wind speed and direction at multiple heights without requiring a physical tower structure. This data collection process, combined with sophisticated computational wind flow modeling that accounts for local terrain effects, is essential to accurately predicting a proposed wind farm's likely long-term energy output, a prediction that directly determines the project's financial viability and the terms on which it can secure financing and power purchase agreements. Wind resource assessment has become increasingly sophisticated over the past two decades, incorporating multi-decade historical weather data, climate modeling to account for potential long-term shifts in regional wind patterns, and increasingly granular local terrain and turbine-wake modeling (accounting for the fact that turbines within a wind farm can partially block wind from reaching downwind turbines, requiring careful spacing and layout optimization), improvements that have meaningfully reduced the financial risk associated with wind project development and contributed, alongside the turbine technology and cost improvements discussed in this chapter's main text, to the wind industry's overall maturation into one of the most bankable and predictable segments of the renewable energy industry.

Wind's Contribution to Rural Economic Development

Much as coal plant retirement carries significant local economic consequences discussed in Chapter 2, wind development has become a significant source of rural economic activity in many of the same Great Plains and Midwest communities affected by coal's decline, through land lease payments to farmers and ranchers hosting turbines, property tax revenue supporting local schools and county government, and construction and ongoing maintenance employment. This economic dimension has helped build durable political support for continued wind development in many rural communities, even amid the localized opposition described earlier in this chapter, illustrating that the energy transition's economic winners and losers are not always cleanly divided along simple urban-rural or partisan lines, but instead depend substantially on which specific communities host which specific technologies, a nuance easily lost in more polarized national debates about renewable energy policy.

Chapter 5:

The Battery Storage Revolution

If solar and wind power represent the renewable transition's supply-side story, utility-scale battery energy storage systems (BESS) represent the technology that has made that transition operationally survivable. Battery storage has gone from an expensive, niche technology in the early 2010s to one of the fastest-growing categories of new generating capacity added to the U.S. grid each year, and it is difficult to overstate how thoroughly it has changed the conversation about renewable energy's reliability limitations. Problems that seemed structurally difficult just a decade ago — the duck curve's steep evening ramp, midday solar curtailment, the need for fast-responding frequency regulation — have all found a partial or substantial solution in batteries.

The Technology and the Cost Collapse

The overwhelming majority of utility-scale batteries deployed on the U.S. grid use lithium-ion chemistry, the same basic battery family used in electric vehicles and consumer electronics, scaled up dramatically and packaged into large, climate-controlled enclosures at substations and dedicated storage sites. Battery costs have fallen at a pace comparable to solar's own cost collapse — lithium-ion battery pack costs fell by roughly 80 to 90 percent between the early 2010s and the mid-2020s, driven by the same manufacturing scale, chemistry refinement, and supply chain maturation (heavily influenced by massive electric vehicle battery manufacturing investment, again concentrated substantially in China) that drove down solar costs.

This cost collapse, combined with the Inflation Reduction Act's investment tax credit for standalone storage (a significant policy change from earlier rules that generally required storage to be paired with a renewable generation project to qualify for tax credits), has driven an extraordinary buildout of utility-scale battery capacity. The United States went from a few hundred megawatts of utility-scale battery storage in the early 2010s to tens of gigawatts by the mid-2020s, with the pace of annual additions continuing to accelerate. California and Texas have emerged as the two dominant battery storage markets in the country, for related but distinct reasons explored below.

California: Batteries as the Duck Curve's Answer

CAISO's battery storage fleet grew from a few hundred megawatts in the late 2010s to well over ten gigawatts by the mid-2020s, making it one of the largest battery storage markets in the world. This growth was a direct, deliberate policy and market response to the duck curve problem described in Chapter 3. California batteries have adopted an operating pattern that has become the signature use case for grid-scale storage nationally: charge during the midday hours when solar output is abundant and wholesale prices are low (sometimes negative, when solar oversupply requires generators to effectively pay the grid to take their power rather than curtail), and discharge during the early evening hours, typically from roughly 4 p.m. to 9 p.m., when solar output collapses but residential demand remains high.

The impact has been substantial and measurable. CAISO's own operational data shows batteries becoming one of the largest sources of electricity supply during the critical early evening hours in recent years, in some hours exceeding the output of the state's entire natural gas fleet during that period — a remarkable shift from just a few years earlier, when gas peaker plants were almost the sole resource covering the evening ramp. This has allowed California to reduce its reliance on the least efficient, most expensive, and most polluting gas peaker plants specifically during the hours they were most needed, while also reducing solar curtailment by giving midday solar surplus somewhere useful to go rather than being wasted.

Texas: Batteries as Insurance Against the Next Uri

ERCOT's battery fleet has grown even faster than California's in percentage terms, driven by a combination of factors: extremely favorable conditions for both wind and solar development that create abundant curtailment-driving surplus generation at various points in the day and year, a relatively fast and flexible interconnection and market structure compared to some other regions, and — significantly — a market and political environment shaped by the trauma of Winter Storm Uri. Texas legislators and regulators, seeking to avoid a repeat of the 2021 crisis, have generally supported market reforms and incentive structures that reward fast-responding, reliable capacity, a category into which battery storage fits naturally given its near-instantaneous response time (batteries can go from zero to full output in a fraction of a second, far faster than even the fastest gas turbines).

Texas batteries have demonstrated real value during subsequent extreme weather events, providing rapid-response capacity during tight grid conditions in ways that have measurably reduced the risk of the kind of emergency load shedding that occurred during Uri. It is worth noting, however, that batteries in their current, predominantly four-hour-duration configuration (discussed below) provide only a partial answer to Uri-style events, which involved multi-day generation shortfalls — a duration far beyond what today's typical battery installations can sustain, a limitation explored further at the end of this chapter.

What Batteries Actually Do for the Grid

It is worth cataloguing, specifically, the range of services batteries provide, because their value extends well beyond the simple "store energy, release energy later" function most people intuitively understand.

Energy arbitrage is the most visible function: charging when electricity is cheap (typically during midday solar abundance or overnight wind abundance) and discharging when electricity is expensive (typically during evening demand peaks), capturing the price difference as revenue while providing the grid with power exactly when it is most needed.

Capacity value refers to a battery's contribution to meeting peak demand and reserve margin requirements, similar to the capacity value concept described for traditional generation in Chapter 1. Batteries have increasingly been credited with substantial capacity value in resource adequacy calculations, particularly in regions like California where the evening peak-demand hours align closely with when batteries are guaranteed to be discharging under their normal operating pattern.

Frequency regulation is a service at which batteries excel almost uniquely well. Because battery inverters can adjust their output nearly instantaneously and with extreme precision, batteries have proven to be dramatically more effective and often cheaper providers of frequency regulation than traditional power plants, which must physically adjust fuel flow and turbine speed to change output — a slower and less precise process. Several ISOs, including PJM, have seen batteries capture a large and growing share of the frequency regulation market specifically because of this performance advantage.

Voltage support and grid-forming capability, an increasingly important and technically sophisticated function discussed further in Chapter 10, allows appropriately configured battery inverters to help stabilize grid voltage and, in advanced "grid-forming" configurations, even help establish and maintain grid frequency independently — a function traditionally provided only by the spinning mass of large synchronous generators.

Transmission and distribution deferral is a less visible but economically significant use case in which utilities deploy batteries at specific locations on their distribution or transmission network to relieve local congestion or capacity constraints, delaying or avoiding the need for expensive wire and substation upgrades — an application likely to grow in importance as datacenter interconnection requests strain local distribution infrastructure, a theme picked up again in Chapter 7.

Black start and resilience services, in which batteries paired with appropriate controls can help restart portions of the grid following an outage, or provide backup power to critical facilities during extended outages, a function of growing interest to both utilities and large customers, including datacenter operators concerned about their own reliability.

Four-Hour Batteries and the Long-Duration Storage Gap

The overwhelming majority of utility-scale batteries deployed through the mid-2020s are configured for approximately four hours of discharge duration at full output — a duration well suited to covering the daily evening demand ramp described above, but insufficient to cover a multi-day generation shortfall of the kind that occurred during Winter Storm Uri, or to fully substitute for the seasonal firm capacity value that traditional plants provide. This four-hour standard emerged largely because it represents something close to an economic sweet spot given current lithium-ion battery costs and the specific daily arbitrage and capacity-value opportunities available in most markets — batteries longer than four hours cost proportionally more (more battery cells needed for the same power capacity) without necessarily capturing proportionally more revenue under current market designs.

This has created growing interest in **long-duration energy storage** technologies capable of discharging for eight, twelve, twenty-four, or even multiple days — a category that includes several emerging non-lithium technologies such as iron-air batteries, flow batteries using various chemistries, compressed air energy storage, gravity-based mechanical storage, and thermal storage systems. None of these long-duration technologies has yet achieved the manufacturing scale and cost reduction that lithium-ion batteries achieved over the past decade, and most remain in pilot or early commercial deployment stages as of the mid-2020s. Closing the long-duration storage gap is widely viewed among grid planners and researchers as one of the most important remaining technology challenges for achieving very high (80 percent or greater) renewable energy penetration while maintaining reliability through extended weather events — precisely the kind of multi-day, wide-area events that pose the greatest reliability risk, as Winter Storm Uri demonstrated.

Batteries as the Grid's New Shock Absorber

The cumulative effect of the battery buildout described in this chapter is that storage has become, in a remarkably short period, one of the grid's primary tools for managing the reliability trade-offs created by rapid solar and wind deployment described in Chapters 3 and 4. Batteries do not eliminate the underlying variability of renewable resources, and they do not yet fully substitute for the multi-day firm capacity that coal, gas, nuclear, and hydro plants provide, particularly during extended extreme weather events. But by absorbing midday renewable surplus and delivering it during evening peaks, by providing frequency regulation and voltage support with a speed and precision traditional generators cannot match, and by deferring costly transmission and distribution upgrades at a moment when the grid faces unprecedented new load growth, battery storage has become, alongside natural gas, one of the two resources utilities most consistently reach for when asked how they intend to keep the grid reliable through the demand surge described in the next section of this book — the AI datacenter boom that Part III now turns to in detail.

Battery Safety and the Fire Risk Question

As utility-scale battery deployment has accelerated, public attention has periodically focused on battery fire incidents, most notably a significant fire at a large battery storage facility at Moss Landing in California in early 2025, which burned for an extended period and prompted evacuations of nearby residents, along with renewed public and regulatory scrutiny of battery storage safety standards. Lithium-ion battery fires are difficult to extinguish using conventional firefighting techniques because the battery cells can sustain a self-perpetuating chemical reaction called "thermal runaway," in which a damaged or overheated cell generates enough heat to trigger the same reaction in neighboring cells, potentially propagating through an entire battery installation despite firefighting efforts, and can in some cases reignite hours or days after apparently being extinguished.

In response to incidents like Moss Landing, industry standards organizations and state regulators have strengthened requirements for battery storage system design, including improved fire suppression systems, physical separation between battery modules to reduce thermal runaway propagation risk, more rigorous testing standards for battery cells and modules, and enhanced emergency response planning and community notification requirements for facilities located near residential areas. The battery storage industry, along with fire safety researchers, has generally characterized these incidents as serious but manageable through continued improvement in system design and safety standards, similar to the safety improvement trajectory other now-mature energy technologies (including, notably, the natural gas and nuclear industries) experienced earlier in their own development, though the incidents underscore that battery storage, like every other energy technology discussed in this book, carries genuine safety considerations that deserve continued attention as deployment scales up, rather than being treated as a uniformly risk-free technology exempt from the safety scrutiny applied to more traditional generation sources.

Second-Life Batteries and the Electric Vehicle Connection

An emerging and potentially significant source of lower-cost battery storage capacity involves repurposing electric vehicle batteries that have degraded below the capacity threshold considered acceptable for vehicle use (typically once a battery has lost roughly 20 to 30 percent of its original capacity) but that retain substantial useful life for stationary grid storage applications, which generally have less demanding performance requirements than vehicle applications. As the electric vehicle fleet built up over the 2010s and 2020s begins to age and its batteries reach end-of-vehicle-life in growing numbers over the coming decade, "second-life" battery storage — built from repurposed electric vehicle battery packs rather than batteries manufactured new specifically for stationary storage — could provide a meaningful, lower-cost supplementary source of grid storage capacity, alongside continued growth in purpose-built new battery storage installations. Several pilot and early commercial second-life battery storage projects have been developed by automakers and independent storage developers, though the segment remains considerably smaller than new-battery utility-scale storage deployment as of the mid-2020s, with its future growth substantially dependent on the pace of electric vehicle fleet turnover over the coming years.

Market Design Challenges: Compensating Batteries Fairly

Battery storage's unique operational flexibility — the ability to both consume electricity (while charging) and produce it (while discharging), often within the same operating day — has created genuine challenges for electricity market designers, who built existing wholesale market rules around the assumption that a given resource is either a generator (only ever producing) or a load (only ever consuming), not both. FERC's Order 841, issued in 2018, required ISOs and RTOs to establish market rules that appropriately accommodate storage resources' dual charging-and-discharging capability, and individual ISOs have continued refining their storage-specific market rules in the years since, addressing questions including how to compensate storage for the multiple, sometimes overlapping services it can provide simultaneously (energy arbitrage, capacity value, and frequency regulation, for example), and how to appropriately account for a battery's limited total energy capacity (as opposed to a traditional generator's largely unlimited fuel supply) when determining how much the battery can be relied upon to sustain output over an extended period during a capacity emergency.

These market design questions remain an active area of ongoing regulatory refinement as of the mid-2020s, and getting them right matters considerably for the pace of continued battery deployment, since a battery storage developer's investment decision depends substantially on having clear, predictable market rules that allow the battery's full range of grid services to be appropriately valued and compensated — a market design challenge that did not exist, in the same form, for the more singular, single-purpose traditional generation resources described in Chapter 2, and one that illustrates how thoroughly battery storage's distinctive technical characteristics have required rethinking market structures originally designed around a very different generation mix.

Sodium-Ion and Alternative Battery Chemistries

While lithium-ion chemistry has dominated utility-scale battery deployment through the mid-2020s, as described earlier in this chapter, ongoing research and early commercialization of alternative battery chemistries could meaningfully reshape the storage landscape over the coming decade. Sodium-ion batteries, using sodium rather than lithium as the primary charge-carrying ion, offer a potentially significant cost advantage since sodium is far more abundant and geographically widespread than lithium, whose global supply remains concentrated in a relatively small number of countries, including Australia, Chile, and China, again connecting to the broader supply chain concentration themes discussed in Part V of this book. Sodium-ion batteries generally offer somewhat lower energy density than lithium-ion batteries, meaning they require more physical space and materials to store the same amount of energy, a disadvantage of limited significance for stationary grid storage applications (where available land, unlike in electric vehicles, is typically less constrained) but one that has slowed sodium-ion's adoption in the electric vehicle market, where energy density and vehicle weight matter considerably more. Several Chinese manufacturers have moved earliest and most aggressively into early commercial sodium-ion battery production, continuing the pattern of Chinese manufacturing leadership in battery technology described in Chapter 11, though several U.S. and European companies have also begun pursuing sodium-ion development, motivated in part by the supply chain diversification benefits sodium-ion's more geographically distributed raw material base could offer relative to continued lithium-ion dependence.

Battery Storage Duration Trends: The Shift Toward Longer Four-to-Eight-Hour Systems

Even within the dominant lithium-ion technology category, average battery storage project duration has begun shifting modestly longer over the mid-2020s, with a growing share of new projects designed for six to eight hours of discharge duration rather than the four-hour standard that dominated earlier deployment, reflecting both continued lithium-ion cost declines (which make longer-duration lithium-ion systems more economically competitive than they were previously) and growing market and regulatory recognition of the reliability value of somewhat longer storage duration, particularly in regions like California where the evening demand ramp described earlier in this chapter can, during certain seasons, extend across a longer window than a four-hour battery can fully cover. This gradual shift toward longer lithium-ion duration represents an intermediate step between the four-hour standard described in this chapter's main text and the considerably longer multi-day duration that the alternative long-duration storage technologies discussed in Chapter 10 aim to eventually provide, illustrating a more gradual, continuous technology evolution rather than a sharp break between "short-duration lithium-ion" and "long-duration alternative chemistry" storage categories.

Storage Siting and Community Engagement

As battery storage deployment has accelerated, siting new storage facilities has, similarly to solar and wind development, increasingly required navigating local community engagement and, in some cases, opposition, particularly following the fire safety concerns discussed earlier in this chapter's supplementary material. Storage developers have increasingly prioritized siting projects at or near existing substations and industrial sites, minimizing the need for new transmission interconnection infrastructure while also situating projects away from dense residential areas where fire safety concerns might generate the greatest community opposition. Some states and localities have developed specific zoning and permitting frameworks for battery storage facilities, distinct from the frameworks historically developed for solar and wind projects, reflecting battery storage's distinct siting considerations (a comparatively small physical footprint compared to solar or wind projects of equivalent capacity, but with the fire safety and emergency response planning considerations described earlier in this chapter) — an evolving area of state and local policy that will likely continue maturing as the pace of battery deployment described throughout this chapter continues.

Batteries and the Datacenter Connection Directly

It is worth drawing an explicit connection, at the close of this chapter, between the battery storage technology described throughout this chapter and the AI datacenter demand growth story that occupies Part III of this book, since the two are increasingly intertwined in practice rather than representing entirely separate developments. Some datacenter operators have begun deploying substantial on-site battery storage directly at their facilities, serving multiple purposes simultaneously: providing a fast-responding bridge power source during the brief transition period when utility power is lost and on-site backup generators are starting up (a role batteries perform far more smoothly than generators alone, which typically require a brief interruption before reaching full output); smoothing the datacenter's own load profile to reduce the fast-ramping grid impact described in Chapter 6, particularly the sudden power draw fluctuations that can occur as large AI training workloads pause and resume; and, increasingly, participating directly in the demand flexibility and curtailment arrangements described in Chapters 7 and 8, discharging stored energy to reduce the facility's net draw from the grid during declared system emergencies, allowing the datacenter to fulfill its curtailment obligations without necessarily reducing its actual computing output at all.

This growing overlap between utility-scale grid battery deployment and datacenter-specific on-site battery deployment illustrates a broader pattern this book has emphasized throughout: the same underlying battery technology and cost trends described in this chapter are being applied simultaneously at the grid level (utility-scale storage smoothing renewable variability) and at the individual large-customer level (datacenter on-site storage smoothing both the facility's own load profile and its interaction with grid emergency curtailment), with both applications drawing on the same manufacturing base, cost trends, and, in some cases, even the same underlying supply chains discussed in this chapter and revisited from a geopolitical and manufacturing perspective in Chapter 11. As battery costs continue to decline, this convergence between grid-scale and customer-sited storage applications is likely to deepen further, blurring what has traditionally been a fairly clean conceptual distinction between "utility resources" and "customer resources" in exactly the way this book's broader narrative, emphasizing systems thinking over siloed technology categories, would predict.

PART III
THE DEMAND SHOCK

Chapter 6: The AI Datacenter Load Surge

For roughly fifteen years, from the mid-2000s through the early 2020s, U.S. electricity demand grew hardly at all. Energy efficiency gains in lighting, appliances, industrial motors, and building systems largely offset population growth and economic expansion, producing a national electricity demand curve so flat that utility long-range planners came to treat near-zero load growth as the default assumption. That era ended abruptly, and the primary reason is the subject of this chapter: the emergence of artificial intelligence as an electricity-intensive industry operating at a scale and pace the grid has never previously been asked to accommodate.

Why AI Computing Consumes So Much Electricity

To understand the scale of the load surge, it helps to understand what is actually happening inside an AI datacenter. Training a large AI model — the process of exposing a neural network to enormous quantities of data and adjusting billions or trillions of internal parameters until the model can perform its intended task — requires running specialized computer chips, most commonly graphics processing units (GPUs) originally designed for rendering video game graphics but exceptionally well suited to the parallel mathematical operations AI training requires, continuously at extremely high power draw for weeks or months at a time. A single high-end AI training chip can draw several hundred to over a thousand watts under full load, and a single training cluster links together tens of thousands of these chips, alongside the networking, cooling, and storage infrastructure needed to support them. A large AI training campus can consume electricity at a rate comparable to a mid-sized American city — several hundred megawatts to over a gigawatt of continuous demand for a single facility, a scale that traditional commercial and industrial customers, even large ones, rarely approached.

Beyond training, "inference" — the process of actually running a trained AI model to answer user queries, generate images, or perform other tasks — has grown into an enormous and rapidly expanding source of electricity demand in its own right as AI products have moved from research projects to mass-market consumer and enterprise tools used by hundreds of millions of people daily. Unlike training, which can in principle be scheduled somewhat flexibly, inference demand tracks real-time usage patterns and must generally be available continuously to serve users around the clock and around the world.

A Load Profile Utilities Have Never Planned For

What makes AI datacenter demand particularly challenging for grid planners is not merely its size but its specific characteristics, several of which differ meaningfully from traditional commercial and industrial loads that utility planning departments have decades of experience forecasting and serving.

Scale concentration: A single datacenter campus can represent a load increase equivalent to adding a significant fraction of an entire utility's existing peak demand, all concentrated at one or a handful of physical locations, often in areas without pre-existing heavy industrial infrastructure or transmission capacity sized to serve such a load.

Flatness and continuity: Unlike most commercial buildings, which have pronounced daily and weekly demand cycles (busy during business hours, quiet at night and on weekends), AI datacenters often run at close to constant, near-maximum power draw twenty-four hours a day, seven days a week, since idle computing hardware represents a poor return on the enormous capital invested in it. This flatness means datacenters do not naturally align with the "peak shaving" opportunities utilities have traditionally used to manage large commercial loads, and it means the load must be served at close to its maximum level nearly continuously, rather than only during a handful of peak hours per year, as is more typical for the reserve-margin planning described in Chapter 1.

Fast ramping: AI computing workloads can shift power draw significantly and quickly — for instance, when a large training job pauses momentarily to synchronize data across thousands of chips, aggregate power draw across a datacenter can drop and then spike again within seconds, creating a load-side ramping challenge conceptually similar to, but distinct in origin from, the generation-side ramping challenges renewable energy creates, discussed in Chapters 3 and 4.

Speed of the request pipeline: Perhaps most disruptive to traditional utility planning, technology companies have been requesting datacenter interconnections on a two-to-four-year development timeline — the time it takes to design, permit, and build a datacenter campus and populate it with computing hardware — which is dramatically faster than the timeline utilities have historically needed to plan, permit, and build major new generation and transmission infrastructure, a mismatch explored in depth in the next chapter.

The Numbers: How Big Is the Surge?

Utility and grid operator load forecasts have been revised upward dramatically and repeatedly since roughly 2022. Multiple major utilities serving datacenter-heavy regions — including Dominion Energy in Virginia, Georgia Power in Georgia, American Electric Power across its multi-state footprint, and Duke Energy in the Carolinas — have published long-term load forecasts showing anticipated peak demand growth of tens of percentage points over a single decade, compared to forecasts published just a few years earlier that anticipated near-flat or low-single-digit growth over the same period. PJM Interconnection, whose footprint includes Northern Virginia's enormous concentration of datacenters — an area sometimes called "Data Center Alley" that hosts a very large share of all internet traffic passing through the eastern United States — has repeatedly revised its own regional load forecasts upward, citing datacenter growth as the dominant driver.

Northern Virginia's Loudoun County and surrounding jurisdictions have become the epicenter of this growth in the eastern United States, hosting one of the largest concentrations of datacenter capacity in the world, a position built over roughly two decades as the region developed dense fiber-optic connectivity, but which has accelerated dramatically with the AI boom. Georgia, Texas, Arizona, and a growing list of other states have each seen billions of dollars in announced datacenter investment, drawn by some combination of available land, favorable tax treatment, access to water for cooling, and, increasingly, proximity to available or plannable electricity generation capacity — a siting factor that has itself become a competitive advantage some states and utilities actively market to prospective datacenter developers.

National-level forecasts published by organizations including the Electric Power Research Institute, the Lawrence Berkeley National Laboratory, and various grid operators have estimated that datacenters could plausibly grow from a mid-single-digit percentage of total U.S. electricity consumption in the early 2020s to a share in the range of ten percent or more by the early-to-mid 2030s under aggressive but plausible growth scenarios — a shift that, if realized, would represent one of the most significant changes in the composition of American electricity demand in decades, comparable in magnitude to the historical rise of air conditioning as a major driver of electricity demand in the mid-twentieth century.

Why the Forecasts Keep Changing

It is worth being candid about the genuine uncertainty embedded in these forecasts, both because intellectual honesty demands it and because that uncertainty is itself an important part of the reliability story. AI datacenter demand forecasts have proven unusually volatile even by the standards of long-term utility planning, for several reasons.

First, AI technology and business economics are evolving extremely rapidly, and it remains genuinely unclear how much computing capacity — and therefore electricity — will ultimately be required to serve the AI applications that prove commercially durable, as opposed to speculative capacity built in anticipation of demand that may or may not materialize at the scale currently projected. Second, individual datacenter developers frequently submit interconnection requests to multiple utilities and multiple sites simultaneously as they compare options, a practice sometimes called "phantom load" or speculative queuing, meaning the raw total of announced interconnection requests in a given region can substantially overstate the load that will actually materialize, since a developer who ultimately builds in one location will typically withdraw requests at the sites not chosen. Third, improvements in AI chip efficiency and datacenter cooling technology could reduce the electricity intensity of a given amount of computing power over time, partially offsetting growth in computing demand itself — though historically, efficiency improvements in computing have tended to be outpaced by growth in the sheer volume of computing performed, a pattern sometimes called Jevons' paradox, in which efficiency gains lower the cost of a resource and thereby increase total consumption of it rather than reducing it.

Utilities and grid operators have responded to this uncertainty by developing more rigorous processes for verifying that interconnection requests represent genuine, financially committed projects rather than speculative placeholders, including requirements for developers to post financial deposits, demonstrate site control, and commit to minimum take-or-pay electricity purchase obligations – reforms discussed further in the next chapter, alongside the broader menu of tools utilities are deploying to meet whatever portion of the forecasted demand surge ultimately proves real.

A Genuine Inflection Point

Whatever uncertainty remains in the precise magnitude of future AI datacenter electricity demand, the qualitative shift is not in serious dispute among utility planners, grid operators, or reliability regulators: after roughly fifteen years of flat demand, the United States has entered a period of the fastest sustained electricity demand growth forecast in a generation, concentrated heavily in specific regions, driven substantially by artificial intelligence infrastructure, and arriving on a timeline that is challenging every assumption embedded in how American utilities have historically planned, permitted, and built new generation and transmission capacity. The remainder of Part III examines how utilities are responding to that challenge, and how the rules governing what happens when the grid cannot meet demand in real time are being quietly rewritten as a direct consequence.

Water, Cooling, and the Other Resource Constraint

Electricity is not the only resource constraint AI datacenters have introduced to grid and infrastructure planning; water for cooling has emerged as an equally significant, if less central to this book's focus, constraint in many locations. Traditional datacenter cooling relies substantially on evaporative cooling systems that consume large quantities of water, and the same regions attractive for datacenter development on electricity-availability grounds – including parts of Arizona, Texas, and Georgia – often face their own water scarcity challenges, creating a secondary resource competition that has become an increasingly prominent factor in local community opposition to proposed datacenter projects and in state and local permitting reviews. Some datacenter developers have responded by adopting air-cooling or liquid-cooling technologies that reduce water consumption, in some cases at the cost of somewhat higher electricity consumption for the cooling systems themselves, illustrating a recurring trade-off in datacenter infrastructure design between different resource constraints that engineers and site selectors must balance simultaneously.

This water dimension connects back to the electricity story in a specific way relevant to siting decisions: several of the emerging generation technologies discussed in Part IV, including small modular reactors and enhanced geothermal systems, also require meaningful water resources for their own cooling systems, meaning datacenter developers and utilities considering co-locating new generation directly with new datacenter load, as discussed in Chapter 9, must account for combined water demand from both the generation source and the datacenter itself, not simply electricity availability alone — an added layer of siting complexity beyond the primarily electricity-focused analysis this book emphasizes.

The International Dimension of Datacenter Siting

While this book focuses on U.S. grid reliability, it is worth noting briefly that the AI datacenter demand surge described in this chapter is a global phenomenon, with technology companies pursuing datacenter development in numerous countries beyond the United States, including significant buildouts in parts of Europe, the Middle East (particularly the United Arab Emirates and Saudi Arabia, which have pursued major AI infrastructure investment as part of broader economic diversification strategies), and, as discussed in Part V of this book, China itself. This global competition for datacenter investment has become an explicit factor in some countries' and states' economic development strategies, with jurisdictions competing to offer favorable electricity pricing, streamlined permitting, and tax incentives to attract datacenter investment, similar to the kind of interstate competition for manufacturing investment that has long characterized U.S. economic development policy, now extended to the specific case of AI computing infrastructure. This competitive dynamic has, if anything, added further pressure on U.S. utilities and regulators to find ways to accommodate datacenter growth quickly, out of concern that excessive delay or restrictive policy could push investment toward competing jurisdictions, domestic or international, willing to offer faster and more accommodating terms.

Efficiency Gains and Their Limits: Will AI's Electricity Appetite Moderate?

A genuinely important, and genuinely uncertain, question this chapter has touched on but deserves further discussion is whether ongoing improvements in AI chip and datacenter efficiency might meaningfully moderate the demand growth trajectory described throughout this chapter. AI computing chip manufacturers, led by Nvidia and a growing set of competitors, have continued to improve the computational efficiency of successive chip generations, delivering more computing performance per unit of electricity consumed with each new product generation. Similarly, datacenter cooling and power delivery infrastructure has become more efficient over time, reducing the "overhead" electricity consumption beyond the computing hardware itself, often measured using a metric called power usage effectiveness (PUE), which compares a datacenter's total electricity consumption to the electricity consumed by its computing equipment alone; state-of-the-art datacenters have achieved PUE values approaching the theoretical minimum of 1.0, compared to considerably less efficient older datacenter designs.

However, as this chapter's main discussion noted, historical experience with computing efficiency improvements generally supports Jevons' paradox — the pattern in which efficiency improvements, by lowering the effective cost of a given amount of computing, tend to increase total demand for computing by more than the efficiency gain reduces the electricity intensity per unit of computing, producing a net increase in total electricity consumption rather than a decrease, even as the technology becomes individually more efficient. Whether this historical pattern will hold for AI-specific computing demand over the coming decade remains a genuinely open empirical question, and one that will substantially determine which of the demand growth scenarios described throughout the remainder of this book — and synthesized in Chapter 13 — actually materializes.

The Financial Markets' Perspective on Datacenter Demand

The datacenter-driven demand growth described throughout this chapter has become a subject of intense interest not only to utility planners and grid operators but to financial markets, with equity analysts covering utility stocks increasingly focused on individual utilities' exposure to datacenter load growth as a significant driver of expected earnings growth, and utility credit rating agencies incorporating datacenter concentration risk into their assessments of individual utilities' credit profiles. This financial market attention has, in turn, reinforced utilities' own internal incentives to pursue datacenter load growth aggressively, since utility earnings under traditional cost-of-service regulation are generally tied to the rate base (the value of the utility's infrastructure investment) on which the utility is permitted to earn a regulated rate of return, meaning substantial new capital investment to serve datacenter load growth — new generation, transmission, and distribution infrastructure — translates relatively directly into increased utility earnings, provided that investment is approved by state regulators as prudent and includes appropriate cost-allocation protections for existing customers, of the kind described in Chapter 8's discussion of consumer advocacy concerns.

This financial dynamic has occasionally raised its own set of concerns among consumer advocates and some regulators: because utilities have a direct financial incentive to pursue and win datacenter development within their service territory, there is a theoretical risk that utility resource planning could tilt toward accommodating speculative or overstated datacenter demand forecasts more readily than a purely reliability-and-cost-minimization-focused planning process might otherwise counsel, a concern that has reinforced regulatory interest in the kind of rigorous demand verification and financial-commitment requirements for large-load interconnection described in Chapter 7, intended to ensure utility investment decisions are grounded in genuinely committed load growth rather than speculative or inflated projections that could, if they fail to materialize, leave existing ratepayers responsible for the cost of stranded infrastructure investment.

Comparing AI Datacenter Growth to Historical Load Growth Episodes

It is useful to place the current AI-driven demand surge in historical context by comparing it to previous episodes of significant U.S. electricity demand growth. The rapid adoption of residential air conditioning across the South and Southwest in the mid-twentieth century drove a multi-decade surge in summer peak electricity demand that fundamentally reshaped utility planning and generation investment in those regions, requiring utilities to build substantial new peaking generation capacity specifically to meet a demand pattern (a sharp, weather-driven summer afternoon peak) that had not previously existed at the same scale. The post-World War II industrial and suburban expansion more broadly drove several decades of sustained, high-single-digit-percentage annual electricity demand growth nationally, a pace of growth that, in percentage terms, actually exceeded the growth rates currently forecast for the AI-datacenter-driven surge described in this chapter, though that mid-twentieth-century growth was more evenly distributed across the entire country and across a more gradual multi-decade timeline, rather than the geographically concentrated, compressed two-to-four-year development timeline that characterizes the current datacenter buildout described throughout this chapter.

This historical comparison offers a measure of reassurance — the U.S. grid has successfully absorbed periods of rapid demand growth before — while also underscoring what is genuinely novel about the current episode: not necessarily the aggregate national percentage growth rate, which remains below some historical peak growth periods, but the extreme geographic concentration and compressed timeline of the current growth, concentrated heavily in a relatively small number of counties and utility service territories, arriving on a development timeline that is dramatically faster than the multi-decade timeline over which previous major demand growth episodes, including the air conditioning adoption wave, actually unfolded.

Measuring Datacenter Electricity Intensity in Everyday Terms

To help readers translate the abstract gigawatt figures discussed throughout this chapter into more intuitive terms, it is useful to note that a single one-gigawatt datacenter campus, running continuously, consumes roughly as much electricity annually as several hundred thousand typical American homes combined. Several of the largest individual datacenter campuses announced or under development in the mid-2020s have targeted eventual capacity in the multiple-gigawatt range at a single site — meaning some individual campuses, once fully built out, could eventually approach the electricity consumption of an entire mid-sized U.S. state's residential sector, concentrated at one physical location. Framed this way, it becomes easier to appreciate why utility planners have found the pace and scale of recent interconnection requests so disruptive to their traditional forecasting methods: a single customer request can dwarf the cumulative load growth a utility might otherwise have expected across its entire service territory over a decade of ordinary population and economic growth.

The Talent and Skills Dimension of AI Infrastructure Growth

Beyond electricity itself, the datacenter buildout described throughout this chapter has created substantial demand for a specialized workforce – electrical engineers, mechanical and cooling systems technicians, construction craft labor for the substations and transmission upgrades large interconnections require, and the utility planning and interconnection engineering staff needed to process the surge of large-load applications described throughout this Part. Several utilities have reported needing to expand their own interconnection and transmission planning staff substantially to handle the volume of large-load requests now arriving, a less visible but genuinely important capacity constraint alongside the equipment and manufacturing bottlenecks discussed in Chapter 7. This workforce dimension echoes the nuclear construction workforce challenges discussed in Chapter 9, reinforcing a theme this book has returned to repeatedly: meeting the demand growth described throughout Part III requires not just capital and technology, but a skilled workforce capable of designing, building, and operating the infrastructure involved, a constraint that takes years to address through training and hiring pipelines regardless of how quickly capital and technology decisions themselves can be made.

Chapter 7:

How Utilities Plan to Meet the Surge

Faced with the load growth forecasts described in the previous chapter, utilities, grid operators, technology companies, and regulators have deployed — and continue to develop — a wide range of strategies to expand and adapt the grid quickly enough to serve new demand without compromising reliability for existing customers. No single strategy is sufficient on its own; the emerging approach is best understood as a portfolio of complementary tools deployed simultaneously, each addressing a different piece of the timeline-and-capacity mismatch described in Chapter 6.

Accelerated Gas Turbine Orders — and a Historic Equipment Backlog

The most immediate and widely adopted response has been a surge in orders for new natural gas-fired generating capacity, since gas plants remain the fastest-to-build form of large-scale, firm, dispatchable generation available at commercial scale. Utilities across the country — including major players serving datacenter-heavy regions such as Georgia Power, Duke Energy, and several utilities within the PJM footprint — have announced substantial new gas generation additions specifically justified by datacenter and broader load growth, in some cases reversing prior long-term plans to reduce gas-fired generation as part of decarbonization commitments.

This surge in demand for new gas plants has run headlong into a manufacturing bottleneck: the handful of companies that manufacture large gas turbines globally — principally General Electric (through its GE Vernova spinoff), Siemens Energy, and Mitsubishi Power — had, over the prior decade of slow U.S. demand growth and a global shift toward renewables, scaled back their turbine manufacturing capacity. The sudden surge in orders, not just from U.S. utilities but from datacenter developers and utilities worldwide responding to similar AI-driven demand growth, has produced turbine order backlogs stretching to the end of the decade in some cases, with utilities reporting having to place orders years in advance of a plant's planned in-service date simply to secure a manufacturing slot — a striking reversal from the historical norm in which turbine availability was rarely the limiting factor in a gas plant's development timeline. This bottleneck has become one of the most concrete, physical constraints on how quickly the "build more gas" strategy can actually be executed, regardless of how much utilities and datacenter developers might wish to accelerate it.

Grid-Enhancing Technologies: Getting More Out of the Wires Already in the Ground

Recognizing that building entirely new transmission lines can take the better part of a decade even under favorable conditions, utilities and grid operators have increasingly turned to a category of technologies known as **grid-enhancing technologies (GETs)** that allow existing transmission infrastructure to carry more power safely, without requiring new construction.

Dynamic line rating is perhaps the most significant of these technologies. Transmission lines have traditionally been assigned a fixed, conservative maximum safe carrying capacity (their "rating") based on worst-case assumptions about ambient temperature and wind conditions, since a power line's capacity is limited by how much it can heat up before sagging dangerously close to the ground or vegetation below. In reality, actual weather conditions are very often more favorable than these worst-case assumptions — wind, in particular, cools power lines and allows them to safely carry meaningfully more current. Dynamic line rating systems use real-time weather sensors, or increasingly, weather-model-based forecasting, to continuously recalculate a line's actual safe carrying capacity, which is very often 10 to 30 percent higher than its static rating during typical operating conditions, unlocking additional transfer capacity on existing infrastructure without new construction.

Reconductoring involves replacing the actual conductor (the metal cable itself) on an existing transmission tower with a newer, advanced-composite-core conductor capable of carrying substantially more current at the same or lower sag, allowing a utility to roughly double a line's capacity while reusing the existing towers and right-of-way — dramatically faster and cheaper than building an entirely new transmission line on a new right-of-way, since the most time-consuming part of transmission construction is typically not the physical wire installation but the multi-year process of siting, permitting, and acquiring rights-of-way for new corridors.

Advanced power flow control devices, which can be installed at strategic points on the grid to actively redirect power flow away from congested paths and onto underutilized parallel paths, similarly allowing more power to move through an existing network without new construction.

Independent studies, including analysis by the Department of Energy and various national laboratories, have estimated that widespread deployment of grid-enhancing technologies could unlock tens of gigawatts of additional transfer capacity on the existing U.S. transmission network at a fraction of the cost and construction timeline of new transmission lines, making these technologies an increasingly central part of utility strategies for accommodating datacenter load growth in the near term, even as longer-term new transmission projects proceed through their multi-year development processes in parallel.

Large-Load Interconnection Tariffs and "Bring Your Own Generation"

Perhaps the most significant structural change utilities have made in response to the datacenter demand surge is the development of entirely new categories of interconnection tariffs specifically designed for very large individual loads — a departure from traditional utility tariff structures that were designed around much smaller commercial and industrial customers and did not anticipate a single customer requesting hundreds of megawatts or more of new capacity.

These large-load tariffs, approved or under consideration by regulators in states including Georgia, Ohio, Virginia, and others, and by ISOs including PJM, typically include several novel provisions. **Minimum take, or take-or-pay, commitments** require a prospective large customer to commit to paying for a substantial minimum quantity of electricity capacity regardless of whether the facility ultimately uses it, protecting existing customers from bearing the cost of infrastructure built to serve a datacenter that is later canceled, downsized, or fails to materialize as forecasted — directly addressing the "phantom load" and speculative interconnection request problem described in the previous chapter. **Longer minimum contract terms**, often a decade or more, similarly protect utilities and their existing customers against the risk of stranded infrastructure investment.

Most significantly for this book's purposes, several of these tariffs include provisions requiring large new loads — again, principally datacenters — to either provide their own on-site backup generation capable of supporting a substantial share of their load during grid emergencies, or to accept explicit contractual terms allowing the utility or grid operator to curtail their electricity draw from the grid first, ahead of other customer classes, during declared capacity emergencies. This is not a hypothetical proposal; it has become an active feature of interconnection agreements and tariff filings in multiple jurisdictions, and it forms the subject of the next chapter, which examines in detail how this arrangement is reshaping the sequence of events during a grid emergency — protecting residential and critical industrial customers by placing large, flexible, often self-generating-capable customers first in line for curtailment.

Demand Flexibility Contracts: Asking Datacenters to Bend, Not Just Back Up

A related but distinct strategy involves negotiating **demand flexibility contracts** directly with datacenter operators, under which the datacenter agrees to voluntarily reduce its electricity draw from the grid — by shifting non-time-sensitive computing workloads to other times or locations, drawing on on-site backup generation or batteries, or in some cases briefly pausing certain operations — during periods of system stress, in exchange for more favorable electricity rates or faster interconnection approval. Some hyperscale technology companies have expressed public interest in this kind of arrangement, in part because it can allow a datacenter to be built and begin operating sooner (since it can be granted a "flexible" or "curtailable" interconnection more quickly than a "firm" interconnection requiring a full, uncurtailed capacity commitment from the utility), and in part because it aligns with corporate sustainability goals by allowing the datacenter's load profile to better match the availability of renewable generation.

The technical feasibility of this approach varies considerably depending on the nature of the computing workload involved. AI model training workloads, which can often be paused and resumed without catastrophic consequence (though not without cost, since idle expensive computing hardware represents lost economic value), are considerably more amenable to demand flexibility than real-time AI inference workloads serving live user requests, which generally must remain available continuously. Research from national laboratories and grid operators has suggested that even modest, occasional curtailment of datacenter load – on the order of a few percent of total annual hours – could unlock substantial additional interconnection capacity without requiring datacenters to meaningfully compromise their overall computing throughput, since the hours in which grid capacity is genuinely constrained represent a small fraction of the total hours in a year. This finding has made demand flexibility one of the more promising near-term tools for reconciling rapid datacenter growth with grid reliability, though it remains in a relatively early stage of contractual and technical implementation as of the mid-2020s.

Transmission Buildout and the Permitting Reform Debate

Even as the tools above are deployed to address near-term constraints, utilities, regional transmission organizations, and federal policymakers continue to pursue longer-term new transmission construction, recognizing that grid-enhancing technologies and demand flexibility, however valuable, cannot substitute indefinitely for genuinely new transmission capacity as load growth continues over the coming decade. Several ISOs, including PJM and MISO, have approved multi-billion-dollar transmission expansion plans specifically citing datacenter-driven load growth as a justification, representing some of the largest transmission investment programs undertaken in decades.

These plans have run into the same permitting and siting challenges described in Chapter 4: new transmission lines routinely take the better part of a decade to permit and build, hampered by multi-state jurisdictional review, local opposition, and unresolved cost-allocation disputes over how the costs of new lines should be shared among the utilities and states that benefit from them. This has produced sustained bipartisan interest in federal permitting reform legislation aimed at streamlining environmental review and siting processes for transmission (and, more broadly, energy infrastructure of all kinds), though as of the mid-2020s, comprehensive permitting reform legislation had not been enacted despite multiple legislative attempts, leaving transmission buildout timelines as one of the most significant remaining structural constraints on how quickly the grid can be expanded to meet the demand growth described in this Part — a constraint that becomes especially significant when compared, in Part V of this book, to the pace at which China has been able to build comparable infrastructure under a very different regulatory and political system.

Behind-the-Meter Generation: Datacenters Building Their Own Power Plants

An increasingly significant strategy, distinct from but related to the grid-connected large-load tariffs described in this chapter, involves datacenter developers building generation capacity directly on their own site, connected directly to their datacenter load rather than to the broader grid — an arrangement often called "behind-the-meter" generation. This approach allows a datacenter to begin operating without waiting for a full utility interconnection and the associated interconnection queue delays described in Chapter 3, since behind-the-meter generation serving only the datacenter's own load, without exporting power to or importing power from the broader grid, can in some cases avoid certain interconnection study requirements that would apply to a grid-connected resource, though the specific regulatory treatment varies considerably by state and utility jurisdiction and has itself become a subject of active regulatory debate.

Several large datacenter projects have been proposed or developed with substantial behind-the-meter natural gas generation, and, as discussed in Chapter 9, some prospective small modular reactor projects are explicitly designed around this kind of direct, co-located power supply arrangement. Behind-the-meter generation offers datacenter developers greater control over their electricity supply timeline and can reduce their exposure to grid reliability risk, but it raises its own set of policy questions this book's later chapters touch upon: whether behind-the-meter generation should still contribute to broader grid reliability and cost-sharing obligations (since a datacenter with substantial on-site generation may still rely on the grid for backup power during its own generation's maintenance outages, creating a "free rider" concern that regulators in several states have begun addressing through specific standby service tariffs), and whether large-scale behind-the-meter development could, over time, reduce the growth of grid-connected demand in ways that actually ease some of the interconnection and capacity pressures described throughout this Part, even as it raises other regulatory and cost-allocation questions of its own.

The Role of State Economic Development Incentives

Beyond the utility-level and grid-operator-level strategies described in this chapter's main discussion, state governments have played an active and often underappreciated role in shaping where and how quickly datacenter capacity has developed, through economic development incentive packages that frequently include substantial tax abatements (particularly exemptions from state and local sales tax on the computing equipment purchases that represent a large share of a datacenter's capital cost) alongside the electricity-specific policy tools described elsewhere in this Part. States including Georgia, Virginia, Texas, and Ohio have each developed extensive datacenter-specific economic incentive programs, competing actively for datacenter investment given the substantial property tax revenue, construction employment, and, to a lesser extent, permanent operational employment (datacenters, once built, typically require relatively modest ongoing staffing relative to their capital investment) that major projects can bring to a local community.

This economic development competition has occasionally created tension with the grid reliability and cost-allocation concerns described throughout this Part: a state or utility eager to attract datacenter investment may face pressure to offer more favorable interconnection terms, faster approval timelines, or more generous electricity rate structures than reliability and cost-allocation caution might otherwise counsel, creating a genuine policy tension between economic development goals and the more conservative, risk-averse approach to large-load interconnection that utilities and reliability-focused regulators might otherwise prefer. How individual states resolve this tension — largely through the specific terms of the large-load tariffs and economic development agreements described throughout this chapter — has become one of the more consequential, if less publicly visible, policy battlegrounds shaping the pace and terms of the AI datacenter buildout described throughout this book.

Coordinating Across Jurisdictional Lines

A final structural challenge worth noting is the coordination challenge created when a single datacenter developer proposes projects across multiple utility service territories or even multiple states simultaneously, as many hyperscale technology companies do as part of their broader site-selection process. Because interconnection studies, large-load tariff terms, and economic development incentives vary considerably from one utility and state jurisdiction to another, developers can find themselves navigating a genuinely fragmented approval landscape, in some cases receiving faster or more favorable terms in one jurisdiction than another for functionally similar projects, an outcome that has prompted some interest in more standardized, multi-state approaches to large-load interconnection policy, though achieving such standardization across the fragmented U.S. regulatory landscape described in Chapter 12 has proven difficult in practice, mirroring the broader jurisdictional fragmentation challenges that have historically slowed interregional transmission development in the United States.

Federal Coordination Efforts

Beyond the state and ISO-level reforms described throughout this chapter, federal agencies have played a growing coordinating role in addressing the datacenter-driven demand surge, even without the kind of unified national planning authority China's system provides, as discussed in Part V. The Department of Energy has convened multiple industry and utility working groups specifically focused on datacenter electricity demand and grid reliability, published analysis quantifying the scale and geographic distribution of anticipated datacenter load growth, and administered funding programs supporting grid-enhancing technology deployment, transmission planning studies, and domestic manufacturing capacity for turbines, transformers, and other critical grid equipment facing the supply chain bottlenecks described elsewhere in this chapter. The Federal Energy Regulatory Commission has similarly opened proceedings specifically examining large-load interconnection practices, co-location arrangements between generation and large loads (relevant to the behind-the-meter and datacenter-nuclear co-location arrangements discussed in this chapter and in Chapter 9), and appropriate cost-allocation principles for transmission investment driven by large new loads, reflecting a recognition that some questions raised by the datacenter buildout — particularly those involving interstate transmission and wholesale market design — fall more naturally within federal rather than purely state jurisdiction, even as the bulk of specific interconnection tariff design and approval remains a state-level responsibility, given the fragmented regulatory structure described throughout this book.

Transformer Shortages: A Second Equipment Bottleneck

Beyond the gas turbine manufacturing bottleneck described in this chapter's main text, the electric grid has faced a parallel and, in some respects, equally significant equipment bottleneck affecting large power transformers — the large, custom-engineered devices used at substations to step voltage up or down between transmission and distribution systems, essential to virtually every new generation, transmission, and large-load interconnection project described throughout this book. Large power transformer manufacturing, like large gas turbine manufacturing, is concentrated among a relatively small number of global manufacturers, and lead times for new large transformers stretched to several years in the mid-2020s, driven by a surge in demand from renewable energy interconnection, datacenter interconnection, transmission expansion, and utility efforts to replace an aging transformer fleet (much of the existing U.S. transformer fleet dates to the mid-twentieth century electrification buildout and is approaching or has exceeded its original design life) all occurring simultaneously.

This transformer bottleneck has, if anything, received less public attention than the gas turbine backlog but has proven equally significant in practice, with utilities and datacenter developers alike reporting transformer procurement as a binding constraint on project timelines in numerous cases. The Department of Energy and several states have responded with targeted funding and streamlined procurement support programs aimed at expanding domestic transformer manufacturing capacity, alongside industry efforts to standardize transformer specifications (reducing the custom-engineering burden that has historically extended transformer manufacturing lead times) — efforts that, as with the turbine manufacturing bottleneck, illustrate a recurring theme throughout this book: the pace of the broader energy transition and demand growth response described throughout this Part depends on an entire ecosystem of specialized manufacturing capacity, not simply on the headline generation and storage technologies that receive the most public and policy attention.

Nuclear Uprates and Life Extension as a Near-Term Bridge

Alongside the strategies already discussed in this chapter, several utilities have leaned further on existing nuclear license renewals and power uprates as a comparatively fast, low-construction-risk way to add firm capacity while the longer-lead-time strategies described elsewhere in this chapter mature. The Nuclear Regulatory Commission has continued approving license renewals extending existing reactors' operating lives to eighty years in some cases, well beyond the forty-year original licensing term most plants were initially granted, reflecting both strong safety performance data accumulated across the existing fleet and the reliability value regulators and utilities alike place on retaining already-built, already-paid-for firm nuclear capacity rather than allowing it to retire and attempting to replace it with newly built generation on the much longer timeline new construction requires. This "sweat the existing assets" strategy, while limited in how much additional capacity it can ultimately unlock compared to genuinely new construction, has provided a meaningful, fast-acting complement to the newer strategies discussed throughout this chapter, buying additional years of firm capacity availability while gas turbine backlogs clear and new nuclear and geothermal projects work through their own multi-year development timelines.

Synthesizing the Utility Response

Taken together, the strategies surveyed throughout this chapter — accelerated gas construction working through real manufacturing constraints, grid-enhancing technologies extracting more capacity from existing wires, novel large-load tariffs that shift risk onto the customers driving growth, demand flexibility contracts that treat datacenter load as a partner in reliability rather than purely a burden upon it, and continued transmission buildout pursued in parallel with permitting reform efforts — represent a genuinely multi-pronged response to a genuinely multi-dimensional challenge. No single strategy discussed in this chapter is sufficient in isolation, and utilities that have navigated the datacenter demand surge most successfully to date have generally been those willing to pursue several of these strategies simultaneously, rather than betting the bulk of their response on any single approach, a pattern of diversified, portfolio-based response that echoes the resource diversification themes emphasized throughout Parts I, II, and IV of this book.

Chapter 8:

Load Shedding in the New Hierarchy

Every strategy described in the previous chapter is, in essence, an attempt to avoid ever having to use the tool of last resort: involuntary load shedding, the deliberate, controlled disconnection of customers from the grid because available generation and imported power cannot meet demand. This chapter examines how that tool has traditionally worked, and how it is being quietly but consequentially redesigned in the age of AI datacenters.

How Emergency Curtailment Traditionally Worked

Grid operators have long maintained a graduated sequence of emergency procedures, formalized by NERC and implemented through each ISO's or utility's own emergency operating plans, designed to preserve reliability through escalating levels of system stress while avoiding involuntary outages for as long as possible.

The sequence typically begins with **public appeals for voluntary conservation** – the familiar requests to raise thermostats, delay running major appliances, and reduce nonessential electricity use during a declared conservation period, communicated through media alerts and, increasingly, smartphone push notifications. If conditions continue to tighten, grid operators move to activating **demand response programs**, contractual arrangements with commercial and industrial customers (and, in some regions, aggregated pools of residential customers with smart thermostats or similar enrolled devices) who have agreed in advance to reduce their electricity consumption on short notice in exchange for compensation. Demand response is a voluntary, compensated, and pre-arranged tool, distinct from the involuntary curtailment discussed below, though it occupies a similar position in the emergency escalation sequence, deployed before involuntary measures if at all possible.

If conditions tighten further, operators may implement a **voltage reduction**, deliberately lowering distribution voltage by a small percentage (typically around 3 to 5 percent) across some or all of the service territory — a measure that reduces electricity consumption modestly, particularly for resistive and lighting loads, without customers losing power entirely, though it can affect the performance of some sensitive equipment.

Only if these measures prove insufficient to prevent an imminent, uncontrolled system collapse do operators move to the most severe tool: **involuntary firm load shedding**, commonly implemented through **rotating outages** (sometimes still called by the older term "rolling blackouts") in which the utility disconnects predetermined blocks of customers, typically organized geographically into numbered "load shed blocks" or "circuits," for limited periods (often 15 minutes to an hour, though emergencies like Winter Storm Uri saw far longer and less predictable outage durations due to the scale and duration of the shortfall), then reconnects that block and moves to the next, cycling through blocks to distribute the burden of the shortfall as evenly as possible across the customer base while reducing total system demand by the amount needed to restore balance.

Critically, **this traditional rotating-outage approach makes essentially no distinction between customer types beyond a narrow set of pre-identified critical facilities** — hospitals, water treatment plants, certain emergency services, and similar critical infrastructure are typically placed on protected circuits that are shielded from rotation to the extent feasible. Beyond that narrow protected category, residential neighborhoods, retail businesses, office buildings, and industrial facilities have traditionally been rotated through outages largely in proportion to their share of the utility's overall circuit topology and customer distribution, without any systematic mechanism to protect, say, a family with an elderly resident dependent on medical equipment, ahead of a large industrial customer, or vice versa, except through the limited critical-facilities exemption.

Why This Traditional Approach Is Being Reconsidered

The traditional approach was designed around a customer base in which no single non-critical customer represented an enormous, disproportionate share of total system load. A large factory might represent a meaningful load, but even substantial industrial customers historically represented at most a small single-digit percentage of a utility's total system demand. The datacenter era has broken this assumption. A single large datacenter campus can represent a load on the same order of magnitude as tens of thousands of homes, concentrated at one interconnection point — meaning that, under the traditional rotating-outage framework, curtailing that single datacenter for an hour could free up as much capacity as curtailing an entire large residential district, without requiring that district's residents to lose power at all.

This has created both a technical opportunity and, increasingly, a policy and even moral argument: if a utility's capacity shortfall can be substantially addressed by curtailing one enormous, non-residential customer rather than rotating outages through tens of thousands of homes, doing so is generally viewed as the more equitable and lower-harm approach — provided the large customer in question can tolerate the curtailment without catastrophic consequences of its own, a condition that, notably, many datacenters are unusually well positioned to satisfy, because a substantial share of datacenter facilities already maintain significant on-site backup generation (typically diesel or natural gas generators, and increasingly batteries) originally installed to protect the datacenter's own operations against ordinary utility outages, completely independent of any broader grid emergency. A datacenter curtailed by the grid operator during a capacity emergency can, in many cases, switch to running on its own backup generators for the duration of the event, maintaining its own operations while relieving pressure on the broader grid — a genuinely rare situation in which a large customer's self-interested reliability investment (protecting itself against outages) directly serves the broader grid's reliability interest (avoiding the need to shed residential and other customers) when the two interests are properly aligned through the right contractual and tariff structure.

How the New Hierarchy Is Being Implemented

This logic has moved from academic discussion to active implementation through several concrete mechanisms.

Curtailable and interruptible large-load tariffs, described in the previous chapter, increasingly include explicit provisions specifying that the utility or grid operator may curtail the customer's grid draw, with limited or no advance notice, during declared system emergencies, in exchange for the customer receiving a more favorable electricity rate, faster interconnection approval, or both, compared to a "firm," non-curtailable service option that would cost more and likely take longer to establish given current interconnection queue constraints. Because datacenter developers are often eager to begin operating as quickly as possible, and because many have on-site backup generation that reduces the actual operational cost of accepting curtailable service, a meaningful share of new large-load interconnections in high-growth regions have been structured on this curtailable basis.

ERCOT's Large Flexible Load program represents one of the more developed examples of this approach in practice, establishing formal processes and market participation rules for large, flexible loads — a category explicitly including datacenters, cryptocurrency mining facilities, and hydrogen production facilities — to register their curtailment capability and participate in programs that call on them to reduce consumption during tight grid conditions, in some cases within minutes of notification, in exchange for compensation and, in some proposed structures, more favorable treatment in the interconnection queue.

PJM's capacity interconnection rights framework and related large-load interconnection reforms under active development in the mid-2020s have similarly moved toward structures that distinguish between "firm" capacity commitments (which cost more, take longer to establish, and are protected from curtailment) and various categories of "non-firm" or conditional service that can be curtailed under defined emergency conditions, with datacenters representing the primary category of customer for whom this distinction has become commercially significant, given the scale of the loads involved.

State regulatory proceedings in Georgia, Ohio, Virginia, Indiana, and elsewhere have taken up variations of this same question, generally converging toward a similar conclusion: large new loads, and datacenters specifically, should bear a greater share of the risk and responsibility associated with their own interconnection, including accepting curtailment during emergencies, rather than having that risk socialized across the entire customer base in the form of either higher rates to fund new firm generation capacity or a heightened risk of residential outages during future emergencies.

The Reordered Social Contract

The cumulative effect of these changes is a genuine, if largely unnoticed by the general public, reordering of the grid's emergency response hierarchy. Where the traditional model treated most non-critical customers as roughly interchangeable during a rotating-outage emergency, the emerging model explicitly places very large, flexible loads — chiefly datacenters, many with substantial on-site backup generation capability — first in line for curtailment, ahead of residential neighborhoods and most industrial and commercial customers, who would only face involuntary outages if curtailing large flexible loads proves insufficient to restore system balance.

This reordering serves multiple interests simultaneously, which likely explains why it has advanced relatively smoothly through regulatory processes despite touching on politically sensitive questions about who bears the burden of grid emergencies. Residential customers and their advocates generally favor an approach that reduces their exposure to outages. Utilities favor an approach that allows them to accommodate large new loads without either overbuilding expensive firm capacity that existing customers would otherwise subsidize, or accepting an unacceptable increase in overall system risk. Datacenter developers, for their part, often find curtailable interconnection service acceptable — sometimes even preferable — given that it can be established faster than firm service, and given that many were already planning to install substantial on-site backup generation for their own independent reliability purposes regardless of what the grid required of them.

It is worth noting important limits to this reordering. Not all datacenters maintain adequate on-site backup generation to fully cover extended curtailment periods, and the reliability of backup generation itself — diesel generators require fuel delivery logistics similar to those that failed during Winter Storm Uri, for instance — is not unlimited. Extended, multi-day capacity emergencies could still eventually require curtailing large flexible loads beyond what their backup generation can sustain, and in a sufficiently severe or prolonged emergency, traditional rotating outages affecting residential and industrial customers could still become necessary even after all available large-load curtailment has been exhausted. The new hierarchy changes the sequence and raises the threshold at which residential and industrial customers face involuntary outages; it does not eliminate that risk entirely, particularly for the kind of extreme, extended, wide-area events — winter cold snaps spanning multiple states, for instance — that have historically produced the grid's most severe reliability failures.

Nonetheless, this reordering represents one of the more significant, if quietly implemented, structural changes to American grid emergency operations in decades, and it stands as one of the clearest examples in this book of how the AI datacenter boom is reshaping not just how much electricity the grid must supply, but the actual operational rules governing how that electricity is allocated when supply falls short.

Legal and Contractual Foundations of Curtailment Rights

The ability of a utility or grid operator to curtail a large customer's electricity supply during an emergency, ahead of residential and other customers, rests on a specific legal and contractual foundation that deserves closer examination. Historically, utility service to any customer, once established, has generally been treated under state law and utility tariffs as a firm commitment, subject to curtailment only under narrowly defined emergency circumstances applied on a nondiscriminatory basis across similarly situated customers — a legal framework that made it difficult, under traditional tariff structures, for a utility to single out one specific large customer for preferential early curtailment without a clear, pre-established contractual basis for doing so.

The large-load tariffs and curtailable interconnection agreements described in this chapter and the previous one address this legal requirement by making the curtailment provision an explicit, voluntarily accepted term of the customer's initial service agreement, established at the time the datacenter or other large load first requests interconnection, rather than an after-the-fact, potentially discriminatory decision applied to an existing firm customer. Because the datacenter developer agrees to these curtailment terms as a condition of receiving service — often in exchange for the faster interconnection timeline and more favorable rate treatment described in Chapter 7 — the arrangement satisfies the nondiscrimination principle that has traditionally governed utility service, while still achieving the practical outcome of protecting residential and other firm customers from the impact of the load in question during genuine system emergencies. This legal structuring detail matters because it means the emerging load-shedding hierarchy described in this chapter is not, strictly speaking, a change to how existing customers can be treated during an emergency; it is instead a change in the terms under which entirely new, large customers are granted service in the first place, a distinction that has made the approach considerably more legally and politically tractable to implement than a retroactive change to existing customer protections would have been.

International Precedent: How Other Countries Handle Large-Load Curtailment

The United States is not alone in confronting the challenge of integrating enormous new datacenter loads into grid emergency planning, and it is instructive to note how other countries have approached similar questions. Ireland, which hosts a disproportionately large concentration of European datacenter capacity relative to its population and grid size (driven substantially by favorable tax treatment and its role as a hub for major technology companies' European operations), has implemented some of the most restrictive datacenter grid connection policies in the developed world, including an effective moratorium on new large datacenter grid connections in the Dublin area for a period in the early 2020s, reflecting concerns that continued unrestricted datacenter growth could threaten the broader grid's reliability and complicate the country's separate renewable energy integration goals. Singapore similarly imposed a moratorium on new datacenter development for a period, before reintroducing a more managed approval process incorporating specific energy efficiency and, increasingly, low-carbon power supply requirements for new projects.

These international examples represent a more restrictive regulatory posture than the United States has generally adopted, reflecting both the smaller physical scale and more limited generation expansion flexibility of grids like Ireland's compared to the much larger U.S. grid regions discussed throughout this book, and different national policy priorities regarding the trade-off between attracting datacenter investment and managing grid reliability risk. The U.S. approach described in this chapter — accommodating growth through curtailable tariffs and a reordered emergency hierarchy, rather than through outright moratoria — reflects a distinctly American preference for market-based, negotiated solutions over direct regulatory restriction, consistent with the broader U.S. regulatory philosophy described throughout this book, though it is worth noting that several U.S. utilities and grid operators have, at times, implemented more limited, temporary interconnection pauses in specific constrained areas, suggesting the more restrictive international approach is not entirely foreign to U.S. practice either, simply less commonly and less permanently applied.

Consumer Advocacy Perspectives

It is worth noting that the reordered load-shedding hierarchy described in this chapter, while broadly welcomed by residential consumer advocates as an improvement over the traditional rotating-outage approach, has not been entirely uncontroversial. Some consumer and environmental justice advocates have raised concerns that the benefits of faster interconnection and more favorable curtailable rate treatment extended to large datacenter customers could, if not carefully structured, shift cost burdens onto residential ratepayers in ways separate from the direct curtailment-sequencing question this chapter focuses on — for instance, if new transmission or generation investment driven by datacenter growth is not adequately allocated to the datacenter customers causing that growth, residential customers could face higher rates even while facing a reduced risk of involuntary outages, an outcome consumer advocates argue would represent an incomplete, rather than a fully successful, resolution of the underlying tension between rapid datacenter growth and protecting existing customers. These cost-allocation concerns, distinct from but related to the curtailment-sequencing focus of this chapter, have become an increasingly prominent feature of state regulatory proceedings addressing large-load tariffs, with several state commissions explicitly requiring that new infrastructure costs attributable to datacenter growth be allocated substantially to the datacenter customers driving that growth, rather than being broadly socialized across the entire residential and commercial customer base, a principle increasingly, though not yet universally, embedded in the large-load tariff structures described throughout this Part.

Technical Requirements for Reliable Curtailment

Successfully implementing the curtailment-based load-shedding hierarchy described in this chapter requires more than a contractual agreement; it requires the underlying technical infrastructure to actually execute rapid, reliable curtailment when called upon. This includes advanced metering and communication infrastructure allowing the utility or grid operator to send a curtailment signal to a specific large-load customer and verify compliance in near-real time, protective relay and control systems at the customer's facility capable of actually reducing electricity draw (or seamlessly transferring to on-site backup generation) within the timeframe the emergency requires, often minutes rather than hours, and robust testing and drill protocols to verify that a given large-load customer's curtailment capability will actually function as contracted when a genuine emergency occurs, rather than discovering gaps in the arrangement only during an actual crisis, when it is too late to correct them.

Several ISOs, including ERCOT through its Large Flexible Load program described in this chapter's main text, have developed specific technical and testing requirements for large loads seeking to participate in curtailable interconnection or demand response arrangements, generally requiring periodic compliance testing and imposing financial penalties for failure to curtail as contracted during an actual emergency event, a mechanism designed to ensure the reordered load-shedding hierarchy described in this chapter functions as a reliable engineering system, not merely a paper agreement, when the grid genuinely needs it.

Communication and Public Trust

A final, less technical but genuinely important dimension of the load-shedding hierarchy changes described in this chapter involves public communication and trust. Because the traditional rotating-outage system, whatever its flaws, was at least broadly familiar and, in its basic logic, comprehensible to the public — outages are shared roughly proportionally across a community — the more customized, datacenter-first curtailment approach described in this chapter requires clear public communication to build and maintain public trust, particularly if residential customers are to understand and accept that a specific class of large customer is being curtailed first specifically to protect their own service reliability. Utilities and grid operators implementing these changes have generally paired the underlying tariff and contractual reforms with public communication efforts explaining the rationale, though public awareness and understanding of these specific reforms remains considerably lower, based on available survey and public comment data from regulatory proceedings, than public awareness of the more visible datacenter growth and electricity cost concerns discussed throughout this book. Building greater public understanding of exactly the kind of reform described in this chapter – one that, properly implemented and communicated, genuinely serves residential customers' interests by shielding them from outage risk – represents an ongoing communications and public engagement challenge for utilities and regulators as the datacenter buildout, and the reordered emergency protocols responding to it, continue to unfold over the coming years.

A Worked Illustration of the New Hierarchy in Action

To make the reordered hierarchy described throughout this chapter fully concrete, consider a plausible illustrative sequence of events during a severe summer heat wave in a PJM-footprint state hosting substantial datacenter load. As afternoon temperatures climb and air conditioning demand surges across the region, PJM's real-time reserve margin begins to tighten. PJM first issues a conservation appeal, asking residential and commercial customers to raise thermostats and delay major appliance use. If conditions continue to tighten, PJM activates its enrolled demand response resources, drawing on commercial and industrial customers who have contracted to reduce load on short notice. If the shortfall persists, PJM (or the local utility, depending on the specific arrangement) issues curtailment notices to large flexible loads under curtailable interconnection agreements – chiefly datacenters with contractual curtailment obligations – which then either reduce their grid draw by shifting to on-site backup generation and batteries, or, where technically feasible, by pausing non-time-sensitive computing workloads. Only if this sequence of measures proves insufficient to restore balance would the utility move to voltage reduction and, as an absolute last resort, rotating outages affecting residential and general commercial and industrial customers.

In this illustrative sequence, the datacenter curtailment step occurs before any residential customer experiences so much as a voltage reduction, let alone an outage — a sequencing made possible entirely by the contractual and tariff reforms described throughout this chapter, and one that would not have been available under the traditional load-shedding framework described at this chapter's opening, in which no formal mechanism existed to single out large flexible loads for earlier-stage curtailment ahead of the broader customer base.

PART IV

THE NEXT GENERATION OF SUPPLY

Chapter 9:

Small Modular Reactors and the Nuclear Renaissance

If natural gas turbine backlogs and transmission permitting delays represent the near-term constraints on meeting AI-driven demand growth described in Part III, small modular reactors (SMRs) represent the technology many utilities, technology companies, and policymakers hope will provide a longer-term answer — a source of firm, carbon-free, always-available electricity that could, in principle, be built faster and more predictably than the large, one-of-a-kind nuclear megaprojects that have defined the industry's troubled recent construction history in the United States.

Why Small Modular Reactors, and Why Now

Conventional large nuclear reactors, typically in the range of 1,000 to 1,600 megawatts of capacity, are essentially custom-engineered, site-built megaprojects. Every large reactor construction project involves enormous quantities of site-specific concrete pouring, piping, and wiring, performed by a workforce that must be assembled and trained for each individual project, with limited ability to repeat and refine construction techniques across multiple builds. This approach has produced some of the most notorious cost and schedule overruns in American infrastructure history: the two-unit Vogtle nuclear expansion in Georgia, the only large new nuclear reactors completed in the United States in the past several decades, took roughly fifteen years from initial construction start to full completion and cost more than double its original budget, with the final price tag exceeding $30 billion for the two units combined.

Small modular reactors are designed around a fundamentally different premise: rather than building a unique, site-specific plant, SMR designs — generally ranging from roughly 50 to 300 megawatts of capacity per module, a fraction of a conventional large reactor's size — are intended to be manufactured substantially in factories, using standardized components and repeatable assembly processes, then transported to a site for final assembly and commissioning, much closer to how large industrial equipment or even modular buildings are manufactured today. Proponents argue this approach can capture the kind of manufacturing learning curve and cost reduction that has driven down costs in industries from automobiles to solar panels, rather than repeating the essentially bespoke, one-off cost structure of traditional nuclear construction. SMRs are also designed with passive safety features intended to reduce the reactor's dependence on active mechanical or electrical systems (such as pumps requiring continuous power) to maintain safe conditions during an emergency, instead relying on natural physical processes like convection and gravity — a design philosophy informed substantially by lessons learned from the 2011 Fukushima disaster.

The Major SMR Developers and Their Approaches

Several companies have emerged as leaders in the U.S. SMR development race, each pursuing a somewhat different reactor design and commercialization strategy.

NuScale Power was among the first SMR developers to receive design certification from the U.S. Nuclear Regulatory Commission, a major regulatory milestone, for its light-water-based modular design. NuScale's proposed first U.S. commercial project, a multi-module plant intended for a site in Idaho serving a consortium of municipal utilities, was canceled in 2023 due to rising cost estimates and insufficient subscriber commitment — a setback that underscored the continued difficulty of achieving cost and schedule discipline even with a design intended to avoid the pitfalls of traditional large-reactor construction, though NuScale has continued pursuing other domestic and international projects since.

X-energy is developing a high-temperature gas-cooled reactor design using a distinctive fuel form called TRISO (tri-structural isotropic) particle fuel, which encapsulates nuclear fuel in layers of ceramic and carbon materials designed to contain radioactive material even under extreme accident conditions. X-energy has partnered with Amazon on plans to develop SMR capacity specifically to serve Amazon's growing datacenter electricity needs, alongside a separate agreement to develop reactors in Washington State in partnership with Energy Northwest, a public power consortium.

Kairos Power is developing a distinctive reactor design using molten fluoride salt as a coolant rather than water, an approach proponents argue offers safety and efficiency advantages, and has signed a substantial agreement with Google to develop multiple SMR units intended to help power Google's datacenter operations, alongside a demonstration reactor project in Tennessee intended to validate the technology at smaller scale before full commercial deployment.

TerraPower, founded with backing from Bill Gates, is developing the Natrium reactor design, which uses liquid sodium as a coolant and incorporates a molten salt thermal energy storage system allowing the plant to boost its electricity output for several hours during periods of peak demand — an unusual design feature explicitly intended to help a nuclear plant complement, rather than compete with, variable renewable generation, echoing the same duck-curve and evening-ramp challenges described in Chapter 3. TerraPower's first demonstration project is under construction in Wyoming, at the site of a retiring coal plant, an arrangement intended to reuse existing transmission interconnection infrastructure and provide continued employment for the local workforce transitioning away from coal.

The Datacenter-Nuclear Connection

The most striking recent development in the SMR and broader nuclear story is the emergence of direct financial and contractual relationships between AI datacenter operators and nuclear generation, described briefly in Chapter 2 and worth examining more fully here. Microsoft's agreement with Constellation Energy to restart a unit at the former Three Mile Island nuclear site in Pennsylvania — rebranded the Crane Clean Energy Center — represents the first commitment to bring a permanently shuttered U.S. commercial nuclear reactor back into service, under a roughly twenty-year power purchase agreement specifically structured to supply Microsoft's datacenter electricity needs. Amazon's agreement with Talen Energy involves both a long-term power purchase arrangement tied to the existing Susquehanna nuclear plant in Pennsylvania and, separately, the SMR development partnership with X-energy described above. Google's agreements with Kairos Power and with utility Constellation Energy (for capacity from Illinois nuclear plants) similarly reflect a broader pattern of hyperscale technology companies seeking to lock in long-term, carbon-free, firm electricity supply directly from nuclear generation, in some cases financing new capacity that would not otherwise have been built or restarted.

This pattern reflects a genuine alignment of interests: AI datacenter operators need enormous quantities of firm, continuously available, ideally carbon-free electricity to meet both their operational needs and their corporate sustainability commitments, and nuclear power is one of the only generation sources that can satisfy all three requirements simultaneously (solar and wind cannot provide the continuous, weather-independent availability datacenters typically require without substantial storage or backup, and natural gas, while firm and available, does not satisfy carbon-free commitments many technology companies have publicly made). Nuclear plants, for their part, have historically struggled to secure the kind of long-term, high-value power purchase commitments that justify their high upfront capital costs, particularly in deregulated wholesale markets where nuclear plants must compete directly against cheaper gas and renewable generation on an hour-by-hour basis; a long-term contract with a creditworthy technology company willing to pay a premium for guaranteed carbon-free, firm power addresses exactly this problem.

Licensing Timelines Versus Demand Timelines

Despite this alignment of interests and the genuine technical promise of SMR designs, a significant gap remains between the timeline on which AI datacenter demand is materializing (the two-to-four-year development cycle described in Chapter 6) and the timeline on which new nuclear capacity, even using modular designs intended to be faster than traditional construction, can realistically be licensed, built, and brought online. Nuclear Regulatory Commission licensing processes, even for designs incorporating substantial standardization, typically require multiple years for design certification and additional years for site-specific licensing, followed by a construction period that, even under optimistic assumptions for first-of-a-kind SMR projects, is likely to span several years beyond that. Most SMR developers' current project timelines point toward first commercial operation in the late 2020s to early 2030s at the earliest, for the first units of each design, with any subsequent manufacturing-scale cost and schedule improvements only materializing across a further multi-year sequence of additional projects.

This means SMRs are best understood, realistically, as a medium-to-long-term answer to the demand growth described in Part III, not a near-term one. The gas turbine additions, grid-enhancing technologies, demand flexibility contracts, and existing nuclear plant restarts and uprates described in Chapters 2 and 7 will necessarily bear the primary burden of meeting demand growth through the remainder of the 2020s, with SMRs and other emerging technologies potentially playing a much larger role in the following decade, provided the current wave of demonstration projects succeeds in proving out both the technology and, just as importantly, the manufacturing and cost discipline that has eluded traditional large-reactor construction in the United States.

How SMRs Could Change Grid Operations

Beyond simply adding generation capacity, widespread SMR deployment could meaningfully change how the grid itself operates, for several reasons explored further in the next chapter and in Part VI's synthesis. SMRs' smaller individual unit size makes co-location directly adjacent to a large datacenter or industrial facility considerably more feasible than for a traditional 1,000-plus-megawatt reactor, potentially allowing a datacenter to receive a substantial share of its power directly from an adjacent plant with minimal reliance on transmission infrastructure — an arrangement that could reduce transmission congestion and interconnection queue pressure, though it raises its own set of regulatory questions about how such "behind the meter" or directly-connected nuclear-to-datacenter arrangements should be treated for reliability, market participation, and cost-allocation purposes, questions state and federal regulators were actively working through as of the mid-2020s.

SMRs' passive safety design and smaller footprint may also make new nuclear siting more feasible at a wider range of locations, including sites like TerraPower's Wyoming project that reuse retiring coal plant infrastructure, potentially easing some of the site-availability and community-acceptance challenges that have historically constrained nuclear development to a relatively small number of locations. Finally, if SMR designs incorporating thermal storage or other flexible-output features (like TerraPower's Natrium design) achieve commercial success, they could provide a source of firm capacity that is also flexible enough to complement variable renewable generation, potentially reducing the grid's dependence on natural gas for the ramping and peaking services described in Chapters 3 and 5 — a development that, if realized at scale, would represent a genuinely significant evolution in how the American grid balances firm and variable resources over the coming decades.

Workforce and Supply Chain Considerations

A frequently underappreciated constraint on the pace of small modular reactor deployment, beyond the regulatory licensing timeline discussed in this chapter's main text, involves the specialized workforce and manufacturing supply chain required to actually build these reactors at scale. Nuclear-grade manufacturing — producing reactor pressure vessels, specialized piping, and other components to the exacting quality and certification standards nuclear applications require — depends on a relatively small number of specialized manufacturers worldwide, a manufacturing base that contracted significantly during the several-decade lull in U.S. nuclear construction between the completion of most of the existing reactor fleet in the 1970s and 1980s and the Vogtle expansion project completed in the 2020s. Rebuilding this manufacturing capacity, along with training the specialized welders, quality assurance inspectors, and construction craft labor that nuclear-grade construction requires, represents its own multi-year undertaking, running in parallel with the regulatory licensing processes described in this chapter's main discussion, and one that several SMR developers have explicitly cited as a key factor in their project timelines and, in some cases, in cost or schedule adjustments to their initial project announcements.

Several SMR developers, and the Department of Energy through various support programs, have pursued deliberate workforce development and supply chain investment initiatives specifically intended to rebuild this nuclear manufacturing and construction capacity ahead of anticipated commercial-scale SMR deployment, drawing in some cases on the existing skilled workforce from the fossil fuel industry (including, notably, workers transitioning from retiring coal plants at sites like TerraPower's Wyoming project) who possess many of the relevant power plant construction and operation skills, even if additional nuclear-specific training and certification is required.

Fuel Supply: The TRISO and HALEU Questions

Several of the advanced SMR designs described in this chapter, including X-energy's TRISO-fueled design and several other developers' designs, require a specific type of nuclear fuel called high-assay low-enriched uranium (HALEU), enriched to a higher concentration than the fuel used in most existing conventional reactors, but still well below weapons-grade enrichment levels. As of the mid-2020s, commercial-scale HALEU production capacity within the United States remained limited, with much of the world's existing HALEU production capacity historically concentrated in Russia — an obviously problematic dependency given the broader geopolitical context following Russia's 2022 invasion of Ukraine and the resulting sanctions and supply chain disruptions affecting numerous Russian-linked commodities and manufacturing capabilities.

The Department of Energy and several private companies have pursued efforts to establish domestic HALEU enrichment capacity specifically to address this supply chain vulnerability, recognizing that a shortage of domestically produced HALEU fuel could become a binding constraint on the pace of SMR deployment even if the reactor licensing and construction challenges described elsewhere in this chapter are successfully resolved. This fuel supply chain question represents another example of a theme recurring throughout this book: successfully deploying a new generation technology at scale typically requires addressing an entire ecosystem of supporting infrastructure and supply chains, not simply the core reactor or power plant technology itself, and delays or bottlenecks in any single link of that broader ecosystem — turbine manufacturing for gas plants, as described in Chapter 7, or HALEU fuel enrichment for advanced SMRs, as described here — can constrain the overall pace of deployment regardless of how successful the core technology itself proves to be.

Public Perception and the Post-Fukushima Nuclear Debate

It would be incomplete to discuss the nuclear renaissance described in this chapter without acknowledging the considerable shift in public and political attitudes toward nuclear power that has occurred over the past decade, a shift that has meaningfully eased (though not eliminated) some of the public acceptance challenges that have historically constrained U.S. nuclear development. Public opinion polling has shown a gradual but genuine increase in public support for nuclear power over the 2010s and 2020s, a shift driven by several factors: growing public concern about climate change and a corresponding recognition of nuclear power's role as a large-scale, carbon-free electricity source; the passage of time and continued strong safety performance of the existing U.S. reactor fleet since the 2011 Fukushima disaster, which had triggered a considerable, though ultimately temporary, decline in public support; and the emergence of the smaller, passively safe SMR designs described in this chapter, which some polling suggests the public perceives as inherently less risky than the large conventional reactors associated in public memory with historical accidents at Three Mile Island, Chernobyl, and Fukushima, even though the fundamental safety case for SMRs rests on specific engineering design features rather than reactor size alone.

This shifting public perception has translated into a meaningfully more favorable political environment for nuclear power across much of the U.S. political spectrum than existed a decade or two earlier, with both major U.S. political parties expressing general support for continued nuclear development in recent years, even as they may disagree on other aspects of energy and climate policy — a rare point of bipartisan alignment that has helped support the federal licensing reforms, funding programs, and general policy environment that has enabled the datacenter-nuclear partnerships and SMR demonstration projects described throughout this chapter to move forward as quickly as they have, even against the backdrop of the licensing timeline and supply chain constraints also described in this supplementary discussion.

International SMR Competition

While this chapter has focused primarily on U.S. small modular reactor developers, it is worth briefly noting that SMR development is a genuinely international competitive field, with significant programs underway in Canada (where Ontario Power Generation has advanced plans for GE Hitachi's BWRX-300 SMR design at its Darlington site, among the furthest-advanced Western SMR projects as of the mid-2020s), the United Kingdom (which has pursued its own SMR competition and development program, including designs from Rolls-Royce), and, as discussed in Chapter 11, China, which has moved its own advanced reactor demonstration projects forward on a broadly comparable timeline to Western developers, backed by its characteristically more centralized state financing and planning model. This international competitive landscape matters for U.S. policy in two respects: it means the United States is not the only country racing to commercialize this technology, creating both competitive pressure and the potential for useful technology and regulatory-experience sharing among close allies, particularly Canada and the United Kingdom, with which the United States has pursued formal nuclear regulatory cooperation agreements intended to reduce duplicative licensing review across jurisdictions for shared reactor designs; and it means that if U.S. SMR development falters or proceeds too slowly, the United

States risks ceding technological and commercial leadership in this specific advanced nuclear technology category to allied nations or, more strategically concerning, to China, a dynamic that has reinforced continued federal policy support for domestic SMR development as a matter of both energy and broader strategic technology competitiveness policy.

Financing Structures for Advanced Nuclear Projects

The nuclear-datacenter power purchase agreements described in this chapter represent one significant financing innovation, but they are not the only mechanism supporting advanced nuclear development. Federal loan guarantee programs administered by the Department of Energy have provided substantial financing support for both the existing Vogtle project (discussed at the start of this chapter) and several SMR demonstration projects, reducing developers' cost of capital by providing federal government backing for loans that private lenders might otherwise view as too risky given the historical track record of nuclear construction cost and schedule overruns described earlier in this chapter. The Inflation Reduction Act's nuclear-specific tax credits, including a production tax credit for existing nuclear generation (helping address the market economics challenge described in Chapter 2's supplementary discussion) and investment tax credits applicable to new nuclear construction, including SMRs, have further improved the financing environment for both existing and new nuclear projects. Some SMR developers have also pursued more novel financing structures, including the kind of direct corporate power purchase and development-cost-sharing agreements with technology companies described earlier in this chapter, which in some cases involve the technology company directly funding a portion of the reactor's development and construction costs in exchange for priority access to the resulting electricity output — a financing innovation that, if it continues to prove successful across the current wave of demonstration projects, could provide a durable, non-federal financing model helping to support continued SMR commercialization beyond the current first generation of demonstration projects.

Cost Trajectories: Will SMRs Actually Get Cheaper?

The central economic bet underlying the small modular reactor development described throughout this chapter is that factory-based, repeatable manufacturing will produce the kind of learning-curve cost decline that solar panels, wind turbines, and batteries have each demonstrated over the past two decades, described respectively in Chapters 3, 4, and 5. This bet remains genuinely unproven at the time of this writing: the first units of any new SMR design will necessarily be built without the benefit of the manufacturing scale and process refinement that only comes from repeated production, meaning first-of-a-kind SMR projects are likely to cost considerably more per unit of capacity than later, more mature units of the same design, a pattern already visible in the cost estimates published for the demonstration projects described earlier in this chapter. Whether subsequent orders achieve the cost reductions SMR developers and their technology-company customers are counting on will depend on securing a sufficient volume of repeat orders to justify sustained factory investment, precisely the kind of sustained order pipeline that has, to date, been difficult for the U.S. nuclear industry to achieve given the multi-decade gap in nuclear construction activity discussed elsewhere in this chapter. The success or failure of this specific cost-reduction bet, over the coming decade, will substantially determine whether small modular reactors become a genuinely significant contributor to the American generation mix, or remain a comparatively niche technology serving a small number of specialized, well-financed corporate customers such as the technology companies discussed throughout this chapter.

Chapter 10:

Other Emerging Sources

Small modular reactors are the most prominent emerging generation technology relevant to the AI-driven demand growth story, but they are not the only one under active development. This chapter surveys several additional technologies – some already reaching early commercial deployment, others still primarily in the research and demonstration stage – and examines a distinct but related technical challenge, grid-forming inverters, that will shape how any of these new resources actually integrate into a renewable-heavy grid.

Enhanced Geothermal Systems

Traditional geothermal power generation has long been limited to a small number of locations with naturally occurring combinations of underground heat, water, and permeable rock – conditions found in the United States primarily in parts of California, Nevada, and a few other western states. **Enhanced geothermal systems (EGS)** aim to dramatically expand geothermal power's geographic reach by artificially creating the conditions traditional geothermal requires: drilling deep wells into hot, dry rock formations found in many more locations, then using techniques adapted substantially from the hydraulic fracturing methods developed by the oil and gas industry to create artificial fracture networks, through which water can be circulated to absorb heat from the rock and be brought to the surface to generate electricity.

Several companies, most notably Fervo Energy, have demonstrated commercial-scale enhanced geothermal projects in Nevada and Utah in the mid-2020s, reporting drilling and well-completion costs that have fallen substantially as the companies have applied lessons and technology directly borrowed from shale oil and gas drilling, an industry that achieved enormous cost reductions of its own over the preceding two decades. Enhanced geothermal offers an attractive combination of attributes for the AI datacenter era specifically: like nuclear power, it provides firm, continuous, weather-independent electricity output, but with a substantially shorter development timeline than nuclear power (geothermal wells can be drilled and completed in months rather than years) and without nuclear power's regulatory and public-acceptance complexities. Several technology companies, including Google, have signed power purchase agreements for enhanced geothermal capacity specifically to serve datacenter electricity needs, viewing it as a nearer-term complement to the longer-term SMR agreements described in the previous chapter. Enhanced geothermal remains a considerably smaller contributor to the national generation mix than solar, wind, or nuclear as of the mid-2020s, but its combination of firm output and relatively fast development timeline has made it one of the more closely watched emerging technologies in the current environment.

Long-Duration Storage Chemistries

Chapter 5 introduced the long-duration energy storage gap — the fact that most utility-scale batteries deployed through the mid-2020s are configured for roughly four hours of discharge, insufficient to cover multi-day generation shortfalls of the kind experienced during Winter Storm Uri. Several companies are developing alternative storage chemistries specifically targeting this gap.

Iron-air batteries, championed most prominently by the company Form Energy, use a reversible rusting reaction (iron reacting with oxygen and reverting back) to store and release energy at a fraction of the cost per unit of energy (though at higher cost per unit of power) compared to lithium-ion batteries, making them potentially well suited to applications requiring many hours to multiple days of discharge duration, even though they cannot discharge or recharge as quickly as lithium-ion systems. Form Energy has begun construction of early commercial-scale projects with several U.S. utilities as of the mid-2020s.

Flow batteries, which store energy in liquid chemical electrolytes held in external tanks rather than within the battery cell itself, offer a design advantage in which energy capacity (tank size) and power capacity (the electrochemical stack) can be scaled independently, making them well suited to longer-duration applications, though flow battery technology has progressed more slowly toward large-scale commercial cost competitiveness than lithium-ion batteries.

Thermal energy storage, including systems that store energy as heat in materials like molten salt, crushed rock, or even specialized concrete blocks, and mechanical storage systems, including gravity-based approaches that raise and lower heavy weights and advanced compressed-air energy storage, round out a diverse field of long-duration storage technologies still working toward the manufacturing scale and cost reduction that lithium-ion batteries achieved over the preceding decade. None of these technologies had reached the deployment scale of lithium-ion storage as of the mid-2020s, but continued cost reduction and successful early commercial projects could meaningfully close the long-duration storage gap over the coming decade, an outcome that would substantially ease the reliability challenges associated with very high renewable energy penetration described throughout Part II of this book.

Fusion's Early Promise — and Its Realistic Timeline

Nuclear fusion – the process that powers the sun, in which light atomic nuclei combine under extreme heat and pressure to form heavier nuclei, releasing enormous energy without the long-lived radioactive waste associated with nuclear fission – has attracted a substantial wave of private investment in the 2020s, with numerous well-funded startup companies pursuing distinct technical approaches, alongside continued government-funded research including the National Ignition Facility, which achieved a milestone "ignition" result (producing more fusion energy output than the laser energy directly delivered to the fuel target) in 2022, and the international ITER project under construction in France.

Fusion's appeal for the AI-driven demand growth story described in this book is obvious: a firm, continuous, carbon-free power source using abundant fuel and producing no long-lived radioactive waste would represent close to an ideal generation technology. However, it is important for readers to understand that fusion power remains, as of the mid-2020s, in a considerably earlier stage of development than any other technology discussed in this chapter or the previous one. No fusion power plant has yet demonstrated sustained, net-positive electricity generation delivered to a grid, and most credible industry and independent expert timelines place the first genuinely commercial fusion power plants in the 2035-to-2040s range at the earliest, with considerable uncertainty even around that estimate given fusion's decades-long history of timelines slipping further into the future. Readers should treat fusion power as a genuinely important long-term research and development priority, worth continued public and private investment, but not as a technology likely to meaningfully contribute to meeting the AI datacenter demand growth described in this book over the next decade.

Grid-Forming Inverters: The Quiet Technology Underlying Everything Else

This chapter closes with a technical concept introduced briefly in Chapter 1 that deserves fuller explanation, because it underlies the successful integration of nearly every emerging technology discussed in this Part, as well as the existing renewable and storage resources discussed in Part II.

Traditional power plants — coal, gas, nuclear, and hydroelectric alike — generate electricity using large rotating machines called synchronous generators, in which a spinning rotor, driven by steam or falling water, physically rotates in sync with the grid's 60-hertz alternating current frequency. This spinning mass provides a valuable and, historically, essentially free byproduct: **inertia**, the physical resistance to sudden changes in frequency that comes from the sheer rotational momentum of all those spinning generators operating in synchronization across the grid. When a large power plant unexpectedly trips offline, grid frequency begins to drop, but the collective inertia of all the other spinning generators still online slows that drop, buying grid operators the crucial few seconds needed for automatic and manual reserve responses to kick in and restore balance before frequency deviates dangerously far from its target.

Solar panels and most wind turbines, along with battery storage systems, do not have this kind of directly grid-synchronized spinning mass; they connect to the grid through power electronics called inverters, which can very precisely control the power the resource delivers but do not inherently provide the same physical inertia a spinning generator provides. As these "inverter-based resources" have grown to represent a larger and larger share of total generation on grids like California's and, increasingly, Texas's, grid operators and researchers have grown concerned about maintaining adequate system inertia, particularly in regions or moments when traditional synchronous generation output is low (for instance, a sunny, windy day with abundant renewable output and correspondingly less traditional generation running) and a sudden large generation loss could produce a dangerously rapid frequency decline that the reduced inertia on the system cannot adequately slow.

The solution increasingly being deployed is the **grid-forming inverter**, a more sophisticated inverter control technology — used primarily in battery storage systems, though increasingly explored for wind and solar as well — that can actively help establish and stabilize grid voltage and frequency, in some configurations even providing a form of "synthetic inertia" that mimics some of the stabilizing behavior traditional synchronous generators provide, rather than simply following a grid frequency and voltage established by other resources (the more common, older "grid-following" inverter design). Grid-forming inverter technology remains an active area of research, standards development, and early commercial deployment as of the mid-2020s, but it is increasingly viewed by grid engineers as an essential technology for enabling continued high renewable and storage penetration without compromising the frequency stability the grid has traditionally taken for granted from its fleet of large, spinning, synchronous power plants — a genuinely foundational technical challenge underlying the entire renewable and storage transition described throughout this book, even though it rarely receives the public attention given to the more visible generation technologies discussed in the surrounding chapters.

Hydrogen as an Energy Storage and Generation Pathway

Beyond the technologies discussed in this chapter's main text, hydrogen has attracted significant interest as a potential long-duration energy storage medium and as a fuel for electricity generation, though it remains, like fusion power, at a comparatively early stage of grid-relevant commercial deployment. The basic concept involves using surplus renewable electricity, during periods of abundant solar or wind output, to produce hydrogen through electrolysis (splitting water into hydrogen and oxygen using an electric current), storing that hydrogen, and later using it either directly as a fuel in modified natural gas turbines (several turbine manufacturers, including GE Vernova and Siemens Energy, have developed and demonstrated turbines capable of burning hydrogen blended with natural gas, and in some cases pure hydrogen) or in fuel cells that convert the stored hydrogen back into electricity through an electrochemical process similar in some respects to a battery, though using an external fuel supply rather than an internal, fixed-capacity electrochemical cell.

Hydrogen's appeal for long-duration storage stems from the fact that, unlike batteries, hydrogen storage capacity (the size of the storage tank or underground cavern used to hold the hydrogen) and hydrogen power generation capacity (the size of the turbine or fuel cell used to convert it back to electricity) can be scaled largely independently, similar to the flow battery advantage described in this chapter's main discussion, potentially allowing hydrogen-based systems to economically provide the multi-day or even seasonal storage duration that lithium-ion batteries cannot practically or economically match. However, the overall efficiency of the "round trip" from electricity to hydrogen and back to electricity remains considerably lower than battery storage's round-trip efficiency, meaning a meaningful share of the original electricity is lost in the conversion process, a real economic disadvantage that has limited hydrogen's near-term competitiveness against batteries for shorter-duration applications, even as its potential advantages for genuinely long-duration and seasonal storage applications continue to attract research, demonstration project, and policy interest, including specific federal funding through the Department of Energy's regional clean hydrogen hub program established under the Inflation Reduction Act and related infrastructure legislation.

Virtual Power Plants: Aggregating Distributed Resources

A final emerging concept worth introducing, distinct from the generation and storage technologies discussed elsewhere in this Part, is the **virtual power plant (VPP)** — a coordinated aggregation of many smaller, individually modest distributed energy resources (residential rooftop solar and battery systems, smart water heaters and thermostats, electric vehicle chargers, and small commercial-scale batteries) that, when coordinated together through software and communication systems, can collectively provide grid services comparable to a traditional, centralized power plant, without requiring the construction of any new large, centralized generation facility at all.

Virtual power plant programs have grown steadily in several states, particularly in regions with substantial existing residential solar and battery adoption, allowing utilities to call upon this aggregated distributed capacity during periods of system stress in a manner conceptually similar to the demand response programs described in Chapter 8, but drawing specifically on the generation and storage capability of enrolled distributed resources, rather than simply asking participating customers to reduce consumption. Some VPP programs have demonstrated the ability to aggregate hundreds of megawatts of effective capacity across tens of thousands of individual residential participants, a scale that, while still modest compared to the utility-scale resources that dominate this book's discussion, has grown enough to be explicitly incorporated into some utilities' and ISOs' resource adequacy and reliability planning as a genuine, if still emerging, contributor to overall system capacity. As electric vehicle adoption and residential battery storage deployment continue to grow over the coming decade, virtual power plants represent one of the more promising avenues for turning what might otherwise be viewed purely as additional distributed load (electric vehicle charging, for instance) into a genuine grid reliability asset, illustrating a broader pattern this book has emphasized throughout: the most effective responses to the demand and reliability challenges described in this book typically involve creatively reconfiguring and coordinating existing and emerging resources, in addition to simply building new, large, centralized generation capacity of the kind that has traditionally dominated American grid planning.

Demand-Side Flexibility Beyond Datacenters

While Chapter 7 discussed demand flexibility specifically in the context of datacenter load, it is worth noting in this chapter, alongside the virtual power plant discussion above, that demand-side flexibility more broadly represents an increasingly important complement to the supply-side emerging technologies discussed throughout this Part. Electric vehicle charging, in particular, represents an enormous and rapidly growing source of potentially flexible electricity demand: because most electric vehicles spend the substantial majority of their time parked rather than driving, and because most vehicle charging does not require immediate completion (a vehicle plugged in overnight generally does not need to reach full charge within the first hour of being plugged in), utilities and technology providers have increasingly developed "managed charging" programs that shift the timing of vehicle charging to align with periods of grid surplus (abundant renewable output or low overall demand) rather than allowing charging to begin immediately whenever a vehicle is plugged in, which can otherwise coincide with and exacerbate the evening demand peak described in Chapter 3's duck curve discussion, since many drivers plug in their vehicles upon arriving home in the early evening, precisely when solar output is falling and residential demand is rising.

As electric vehicle adoption continues to grow, managed charging and, more advanced, "vehicle-to-grid" technology (in which a parked electric vehicle's battery can not only draw power from the grid but also discharge power back to the grid or to a home during periods of need, functioning as a mobile, distributed complement to the stationary battery storage described in Chapter 5) represent an increasingly significant potential source of demand-side flexibility and even storage capacity, though vehicle-to-grid technology remains at an earlier stage of commercial deployment than stationary storage, given the additional battery degradation concerns, vehicle warranty considerations, and standardization challenges involved in enabling widespread bidirectional vehicle charging.

Bringing the Emerging Technologies Together

This chapter and the previous one have surveyed a wide range of emerging generation and storage technologies – small modular reactors, enhanced geothermal, iron-air and flow batteries, hydrogen, virtual power plants, and managed electric vehicle charging – each at a different stage of technical maturity and commercial deployment, and each addressing a somewhat different piece of the reliability puzzle this book has examined throughout. It is worth emphasizing, in closing this chapter, that none of these technologies is likely to single-handedly resolve the reliability challenges described throughout this book; rather, as this book's synthesis in Chapter 13 will argue more fully, the most likely and most resilient path forward involves a genuinely diversified portfolio combining continued traditional generation, continued rapid solar and wind deployment, continued battery storage growth, gradual integration of emerging firm technologies like SMRs and enhanced geothermal as they mature, and growing contributions from demand-side flexibility and aggregated distributed resources – a portfolio approach that mirrors, at the level of specific technology choices, the same fuel-diversity reliability principle introduced in Chapter 2's discussion of the lessons learned from Winter Storm Uri, now applied across the full range of generation, storage, and demand-side technologies surveyed throughout this book.

Advanced Nuclear Fuel Recycling as a Longer-Term Consideration

A final emerging concept worth brief mention, adjacent to the small modular reactor discussion in Chapter 9, involves advanced fuel cycles and used-fuel recycling technologies that some next-generation reactor designs, including several of the molten-salt and sodium-cooled designs mentioned in that chapter, are engineered to eventually utilize. These approaches aim to extract significantly more energy from a given quantity of mined uranium than conventional once-through fuel cycles achieve, while potentially reducing the volume and longevity of high-level nuclear waste requiring long-term storage or disposal, a subject of longstanding public and policy concern surrounding nuclear power generally. Fuel recycling technology remains commercially undeployed at scale in the United States as of the mid-2020s, constrained by both technical maturity and the nonproliferation policy considerations that have historically shaped U.S. reluctance to pursue commercial fuel reprocessing compared to some other nuclear nations, but it remains an active area of Department of Energy research investment and a potential long-term complement to the advanced reactor designs discussed in the previous chapter, illustrating that the nuclear technology story explored throughout Part IV extends beyond reactor design alone to the broader fuel cycle infrastructure that supports it.

PART V

THE GLOBAL RACE

Chapter 11:

China's Generation Build-Out

No account of American grid reliability in the age of artificial intelligence is complete without confronting an uncomfortable comparison: while U.S. utilities navigate interconnection queues, turbine backlogs, and multi-state transmission permitting battles, China has been adding electricity generation capacity — across every major technology category — at a pace that dwarfs anything happening in the United States, executing in a single year what American utilities and regulators might take a decade to permit and build. Understanding this gap is essential to understanding the strategic stakes of the reliability challenges described throughout this book.

The Scale of Chinese Coal Additions

Perhaps the most striking and, to many Western observers, counterintuitive fact about China's electricity system is that even as it has become the world's largest manufacturer and deployer of solar panels, wind turbines, and batteries, it has continued to approve and construct enormous quantities of new coal-fired power plant capacity. Chinese provincial and national authorities approved tens of gigawatts of new coal plant permits in recent years, and China's total coal-fired generating fleet substantially exceeds that of the rest of the world combined. This continued coal expansion reflects a set of priorities distinct from the U.S. and European approach: Chinese energy planners have generally treated new coal capacity as an insurance policy for grid reliability and energy security, a hedge against the variability of the country's own rapidly growing renewable fleet and against the geopolitical risk of dependence on imported fuels, even while simultaneously pursuing the world's most aggressive renewable energy deployment program. This "both/and" approach — massive coal expansion alongside massive renewable expansion —

differs fundamentally from the more binary "coal versus renewables" framing that has often characterized U.S. and European energy policy debates.

The Scale of Chinese Solar and Wind Additions

China's solar and wind deployment pace is, in a word, staggering by comparison to any other country, including the United States. China has, in recent years, installed more new solar capacity in a single year than the United States' entire cumulative installed solar fleet built up over the preceding two decades. China similarly leads the world by a wide margin in annual new wind capacity additions, both onshore and offshore. This deployment pace is underpinned by China's overwhelming dominance of global manufacturing capacity for solar panels, wind turbine components, and lithium-ion batteries — a dominance built through more than a decade of sustained state industrial policy, manufacturing subsidies, and strategic investment that gave Chinese manufacturers decisive scale and cost advantages, the same underlying cost collapse described in Chapters 3 and 5 of this book, but achieved substantially through Chinese manufacturing capacity regardless of where the panels, turbines, and batteries are ultimately installed. China's dominant position in the global supply chains for these technologies means that even U.S. solar and battery deployment, described in earlier chapters as a domestic American success story, remains substantially dependent on components, materials, or manufacturing equipment originating from or influenced by Chinese industrial capacity, a dependency that has become an increasing focus of U.S. trade policy, tariffs, and domestic manufacturing incentives under the Inflation Reduction Act and related legislation, aimed at building a more independent U.S. supply chain, albeit from a starting position of substantial existing Chinese market dominance.

China's Nuclear Program

China's nuclear power program has similarly advanced at a pace far exceeding the United States' recent nuclear construction experience. China has multiple nuclear reactors under simultaneous construction at any given time, compared to the single Vogtle project (described in Chapter 9) that represented the entirety of recent U.S. large-reactor construction. Chinese nuclear construction has achieved meaningfully shorter timelines from construction start to commercial operation than recent U.S. and European nuclear projects, an advantage attributed to sustained construction experience across a continuous pipeline of projects (allowing workforce skills and supply chains to be maintained and refined project to project, rather than atrophying between infrequent projects as has happened in the U.S. nuclear construction industry), standardized reactor designs built repeatedly rather than the more customized approach historically common in Western nuclear construction, and a state-directed regulatory and financing environment that can move projects from approval to construction more quickly than the multi-layered environmental review, public comment, and financing processes typical in the United States. China has also moved more aggressively than the United States into demonstration and early commercial deployment of advanced reactor designs, including its own small modular reactor and high-temperature gas-cooled reactor demonstration projects, running on a broadly comparable technology timeline to the U.S. SMR developers described in Chapter 9, but backed by a state-directed development and financing model rather than the more market-and-venture-capital-driven U.S. approach.

Ultra-High-Voltage Transmission: China's Transmission Advantage

Perhaps the starkest infrastructure contrast between the United States and China lies in transmission. China has built an extensive network of **ultra-high-voltage (UHV)** transmission lines — operating at voltages considerably higher than the highest-voltage transmission lines commonly used in the United States — specifically designed to move enormous quantities of electricity over very long distances, connecting renewable-rich and coal-rich regions in China's west and north to the enormous demand centers along its eastern coast, over distances of a thousand miles or more, with power losses over such long distances kept remarkably low thanks to the higher voltages involved.

This UHV transmission network was built as a deliberate, centrally planned national infrastructure program, executed by China's State Grid Corporation (one of the largest utilities in the world, effectively a state-owned monopoly covering the vast majority of the country) without the multi-state jurisdictional permitting battles, competing private transmission developers, and fragmented cost-allocation disputes that have persistently slowed comparable interregional transmission projects in the United States, as described in Chapters 4 and 7. Where a single major interregional U.S. transmission project can take the better part of a decade or more to permit and build, China has constructed a comparably ambitious nationwide UHV network across a shorter overall timeframe, a difference in execution speed that reflects fundamental structural differences between the two countries' governance and regulatory systems, examined further in the next chapter.

Battery Storage in China

China's battery storage deployment has followed a similar pattern to its solar, wind, and nuclear buildout: rapid, large-scale, and centrally coordinated. Chinese utility-scale battery storage capacity has grown extremely rapidly, driven by the same underlying lithium-ion battery cost collapse described in Chapter 5 — a cost collapse substantially achieved through Chinese battery manufacturing scale in the first place — combined with government mandates in many Chinese provinces requiring new renewable energy projects to be paired with a specified minimum amount of storage capacity, a more directive policy approach than the market-and-tax-credit-driven storage deployment model that has predominated in the United States.

Why the Comparison Matters for This Book

It would be a mistake to read this chapter as an argument that China's state-directed model is straightforwardly "better" than the United States' more market-based, locally regulated approach — the two systems reflect different political values, and China's approach carries its own well-documented costs, including limited public input into siting decisions, environmental and human rights concerns associated with some elements of its energy supply chains, and a degree of centralized economic planning that would be difficult to reconcile with core American political and economic institutions. Nor does rapid capacity addition automatically equate to a more reliable grid in practice; China's grid faces its own reliability challenges, including periodic regional power shortages during extreme heat waves that have required emergency industrial curtailment.

But from the narrow, specific lens of this book's central question — can a country add electricity generation and transmission capacity fast enough to keep pace with a period of extraordinary, AI-driven demand growth — the comparison is instructive and, frankly, sobering. China has demonstrated that when a national government prioritizes rapid electricity infrastructure buildout and is willing to accept the trade-offs (continued heavy coal use, centralized planning, less local input into siting decisions) that rapid buildout requires, adding many multiples of the generation and transmission capacity the United States adds in a comparable period is achievable. The next chapter turns directly to the question this comparison raises for American policymakers, utility planners, and readers of this book: what would it take for the United States to meet its own future load growth at a pace comparable to what China has already demonstrated, and what are the realistic prospects for achieving that pace given the structural differences between the two countries' systems?

China's Grid Structure Compared to the U.S. Model

China's electricity system is dominated by two enormous state-owned grid companies — State Grid Corporation of China, which covers the large majority of the country's geographic area and population, and China Southern Power Grid, which covers several southern provinces — a structure fundamentally different from the fragmented, multi-utility, multi-ISO landscape described in Chapter 1 of this book. These state grid companies operate both transmission and, in most areas, distribution infrastructure as unified, centrally directed entities, coordinating closely with China's national economic planning apparatus, including the National Development and Reform Commission and the National Energy Administration, which set national and provincial generation capacity targets as part of China's broader five-year economic planning cycles.

This structural difference has direct implications for the pace comparison discussed in this chapter: because State Grid and China Southern Power Grid operate with unified planning authority across enormous geographic footprints, and because China's national planning process can set binding capacity and infrastructure targets that provincial and local authorities are expected to help implement, large infrastructure projects — particularly the ultra-high-voltage transmission lines discussed in this chapter's main text — can move from planning approval to construction with a degree of coordinated authority that the fragmented U.S. system, divided among dozens of separate utilities, multiple competing ISOs and RTOs, and fifty separate state regulatory commissions, simply cannot replicate, for reasons rooted in fundamental differences between the two countries' governance structures rather than any single fixable policy or regulatory choice.

Chinese Domestic AI Datacenter Growth and Chip Constraints

China's own domestic AI industry has grown substantially in recent years, driven by major Chinese technology companies including Alibaba, Tencent, Baidu, and a growing set of specialized AI companies, alongside significant state investment in AI research and infrastructure as part of China's explicit national strategic priority to achieve leadership in artificial intelligence technology. However, as noted in this chapter's main discussion, U.S. export controls restricting the sale of the most advanced AI computing chips (particularly high-end graphics processing units manufactured using the most advanced semiconductor fabrication processes, largely produced by Taiwan Semiconductor Manufacturing Company under license from U.S. chip designers including Nvidia) to Chinese customers have created a meaningful, though incomplete and imperfectly enforced, constraint on the scale of Chinese AI computing infrastructure development, at least for AI training workloads that benefit most from the highest-performance available chips.

China has responded to these export controls through several parallel strategies: continued investment in developing domestic semiconductor manufacturing capability intended to eventually reduce dependence on foreign-controlled advanced chip manufacturing, development of Chinese-designed AI chips (including chips developed by Huawei and other domestic firms) that, while generally still lagging the most advanced U.S.-designed chips in raw performance, have continued to improve, and, in some cases, the use of larger quantities of less advanced chips to compensate for individual chip performance limitations, an approach that, notably, tends to increase rather than decrease the total electricity consumption required to achieve a given amount of AI computing capability, since a larger number of less efficient chips consuming electricity collectively can approach the computing output of a smaller number of more efficient advanced chips, but typically at a higher aggregate electricity cost. This dynamic means that, somewhat counterintuitively, U.S. export controls intended to constrain Chinese AI development could, at the margin, increase rather than decrease Chinese AI datacenter electricity demand, an unintended consequence policymakers and analysts have noted as part of the broader, complex relationship between AI technology policy, semiconductor trade policy, and electricity demand growth that spans both the U.S. and Chinese contexts discussed throughout this Part.

Environmental and Human Rights Considerations in Chinese Energy Supply Chains

Any honest treatment of China's generation and manufacturing buildout, of the kind this book aims to provide, must also acknowledge the environmental and human rights concerns that have been raised regarding specific elements of China's energy supply chains, concerns that have become an increasing focus of U.S. and international trade policy in their own right. Significant polysilicon production capacity for solar panel manufacturing has historically been concentrated in China's Xinjiang region, an area where credible human rights organizations, journalists, and governments including the United States have documented forced labor practices affecting the Uyghur ethnic minority population, leading the United States to enact the Uyghur Forced Labor Prevention Act, which restricts the import of goods, including solar components, linked to forced labor in the region, and has required U.S. solar developers and importers to implement more rigorous supply chain verification and, in some cases, to shift sourcing toward manufacturers with verified non-Xinjiang supply chains.

Similarly, China's continued heavy reliance on coal-fired power generation, described earlier in this chapter, carries substantial air quality and greenhouse gas emissions consequences, even as China has simultaneously become the world's dominant manufacturer and deployer of the solar, wind, and battery technologies most central to global decarbonization efforts — a genuine and important tension in China's overall energy and environmental profile that complicates any simple characterization of China's energy system as either straightforwardly a decarbonization leader or a climate liability, and one that deserves acknowledgment alongside the narrower infrastructure-pace comparison that is this Part's primary focus.

The Belt and Road Dimension: Exporting Chinese Grid Technology

China's domestic generation and transmission buildout, described throughout this chapter, has been paralleled by an active strategy of exporting Chinese-manufactured solar panels, wind turbines, batteries, and, increasingly, entire power plant and transmission projects to other countries, substantially through financing and construction arrangements associated with China's Belt and Road Initiative, a global infrastructure development program China has pursued since 2013 across dozens of countries in Asia, Africa, Latin America, and elsewhere. Chinese state-owned and state-linked companies have built or financed a substantial share of new power generation capacity in numerous developing countries over the past decade, including both renewable projects and, notably, coal-fired power plants that China has continued to finance abroad even as international climate finance institutions and many Western governments have moved away from financing new coal capacity internationally, though China itself announced in 2021 that it would stop building new coal-fired power projects abroad, a policy shift that has somewhat reduced, though not eliminated, Chinese-financed coal development in other countries in the years since.

This export dimension matters for the U.S.-China comparison at the heart of this Part for a specific reason: it means China's manufacturing scale advantages in solar, wind, and battery technology, described throughout this chapter, translate not only into faster domestic Chinese deployment but into substantial global market share and influence over energy infrastructure development in numerous other countries, a form of geopolitical and economic influence that compounds the direct domestic infrastructure-pace comparison this Part has emphasized, and one that U.S. policymakers focused on the narrower question of domestic grid reliability, this book's primary focus, should nonetheless understand as part of the broader strategic context in which the domestic reliability challenges described throughout this book are unfolding.

Lessons for U.S. Policy from the China Comparison

Having examined China's buildout in considerable detail across this chapter, it is worth drawing out, in summary form, the specific lessons this comparison offers for U.S. policy, beyond the general observation that China builds infrastructure faster. First, the comparison underscores that manufacturing scale and supply chain control matter enormously for the cost and pace of energy technology deployment, reinforcing the case for continued, patient U.S. investment in domestic manufacturing capacity across solar, battery, and advanced nuclear supply chains, even though such investment will take considerably longer than a single legislative cycle to meaningfully close the gap with China's already-established manufacturing dominance. Second, the comparison underscores that unified, proactive transmission planning — rather than a purely developer-initiated, reactive interconnection queue process — can meaningfully accelerate infrastructure deployment, lending support to continued U.S. efforts toward more proactive, scenario-based transmission planning of the kind some ISOs, including MISO, have begun to pursue through their long-range transmission planning processes, even short of the kind of fully centralized national planning authority China's system provides. Third, and perhaps most importantly, the comparison should inform how U.S. policymakers and the public calibrate their expectations and sense of urgency: the reliability and demand-growth challenges described throughout this book are unfolding on a genuinely rapid timeline, and a U.S. policy and regulatory response calibrated to the more deliberate pace of past decades is unlikely to prove adequate to the moment, a conclusion this chapter's comparison to China illustrates with particular clarity, even if replicating China's specific governance model is neither possible nor desirable within the American constitutional and political system.

Critical Minerals and the Supply Chain Beneath the Buildout

Underlying both the U.S. and Chinese generation buildouts described throughout this Part is a shared dependence on critical minerals — lithium, cobalt, nickel, rare earth elements used in wind turbine magnets and other components, and copper for the vast quantities of wiring every generation and transmission technology in this book requires. China holds a dominant position not only in manufacturing but in the processing and refining of many of these critical minerals, even where the raw ore itself is mined elsewhere, giving Chinese industry substantial influence over global supply chains and pricing for materials essential to nearly every technology discussed in this book, from batteries to wind turbines to the transmission lines connecting them all to the grid. The United States and its allies have pursued a range of initiatives aimed at diversifying critical mineral supply chains, including new domestic mining and processing investment and coordinated sourcing agreements among allied nations, but these efforts, similar to the domestic manufacturing initiatives discussed elsewhere in this book, will require many years to meaningfully reduce the current degree of dependence on Chinese-dominated mineral processing capacity.

Closing Perspective on the Comparison

The China comparison developed across this chapter is not offered as an argument for American emulation of Chinese governance, but as a diagnostic tool: it isolates, with unusual clarity, how much of the pace gap between the two countries' infrastructure buildouts stems from genuinely fixable process and manufacturing constraints, as opposed to more fundamental resource, technology, or capital limitations that no amount of policy reform could fully overcome. That distinction — between fixable friction and fundamental constraint — turns out to favor cautious optimism about the American system's long-term capacity to respond to the demand growth described throughout this book, even as it counsels considerable urgency about the pace of reform still required, a theme the next chapter develops directly.

Chapter 12:

Matching Capacity to Future Load U.S. vs. China

Having examined the scale of China's generation and transmission buildout in the previous chapter, this chapter turns to the comparative question at the heart of Part V: how does China's pace of capacity addition compare to what the United States needs to achieve to meet its own future load growth, and what does that comparison reveal about the structural advantages and disadvantages each country's system brings to the challenge?

Comparing the Drivers of Load Growth

It is worth first noting that the United States and China face meaningfully different drivers of electricity demand growth, even though both countries are experiencing significant growth after periods of slower demand increase. China's electricity demand growth has been driven substantially by continued industrialization, the electrification of transportation (China has achieved by far the world's highest electric vehicle adoption rates), the electrification of heating in some regions previously reliant on coal for direct heat, and a growing but still comparatively smaller contribution from its own rapidly expanding domestic AI and datacenter industry, which has grown considerably but from a smaller base than the U.S. hyperscale datacenter buildout described in Chapter 6, and which has, in recent years, been affected by U.S. export controls on the most advanced AI computing chips, somewhat constraining (though by no means eliminating) the scale of Chinese AI datacenter electricity demand relative to what would likely have occurred absent such controls.

The United States, by contrast, faces electricity demand growth driven predominantly by the AI datacenter buildout described throughout Part III of this book, alongside a meaningful but comparatively smaller contribution from electric vehicle adoption (which has grown steadily but at a considerably slower pace and lower penetration rate than in China) and a modest resurgence in domestic manufacturing investment associated with federal industrial policy, including the CHIPS Act's semiconductor manufacturing incentives and the Inflation Reduction Act's clean energy and manufacturing provisions. In short: China's load growth is broader-based and reflects a still-industrializing, still-electrifying economy at a different stage of development, while the United States' load growth is more narrowly concentrated in a single, extraordinarily fast-growing sector — AI infrastructure — layered on top of an otherwise mature, already highly electrified economy.

Comparing the Pace of Capacity Addition

Despite these differing drivers, the comparison of raw capacity addition pace is stark and directly relevant to this book's reliability thesis. China's annual additions of solar, wind, coal, and nuclear capacity each individually exceed, often by a wide margin, the United States' annual additions of the same technology categories, even after accounting for the substantial acceleration in U.S. solar, wind, and battery deployment described in Part II of this book. China's transmission buildout, executed through its centrally coordinated UHV program described in the previous chapter, has similarly proceeded at a pace that U.S. transmission planners and permitting reform advocates frequently cite, with evident frustration, as a benchmark the American system has structurally struggled to approach.

This gap exists despite the United States possessing what should be, on paper, significant advantages: a highly developed capital market able to finance large infrastructure projects, a technologically sophisticated engineering and manufacturing base, and a wealthy economy able to absorb substantial infrastructure investment. The gap instead reflects structural and institutional differences in how the two countries make and execute infrastructure decisions.

Structural Explanations for the Gap

Several factors, discussed at various points throughout this book, combine to explain why the United States has struggled to match China's pace of capacity addition, and are worth drawing together explicitly here.

Permitting and environmental review complexity. U.S. energy infrastructure projects — whether a new transmission line, a new gas plant, or a new nuclear facility — are subject to multiple layers of environmental review, including federal reviews under the National Environmental Policy Act, state-level permitting processes, and, for transmission specifically, the multi-state jurisdictional approval processes described in Chapter 4. These processes exist to protect legitimate public interests, including environmental protection and local community input, values reflected in core American legal and political traditions that a purely speed-focused comparison to China's system should not casually dismiss. But their cumulative effect has been a substantially slower infrastructure development timeline than China's more centrally directed and streamlined approval process.

Fragmented jurisdiction and cost allocation. The United States' electricity system is governed by a patchwork of federal, state, and regional authorities — the Federal Energy Regulatory Commission, NERC, individual state public utility commissions, and the various ISOs and RTOs described in Chapter 1 — reflecting the country's federalist political structure, in which no single entity has the authority China's State Grid Corporation holds to plan and execute infrastructure investment across the entire country. This fragmentation has repeatedly slowed interregional transmission projects specifically, as described in Chapter 4, because projects benefiting multiple states or utility service territories require agreement among multiple, often competing, jurisdictions on how costs and benefits should be shared.

Manufacturing supply chain dependence. As described in Chapter 11, the United States' solar, wind, and battery deployment — despite being a genuine domestic policy and market success story in its own right — remains substantially dependent on global supply chains in which Chinese manufacturing holds a dominant position, a dependence the Inflation Reduction Act and CHIPS Act have begun to address through domestic manufacturing incentives, but one that will take years to meaningfully rebalance given the scale of existing Chinese manufacturing capacity and the time required to build comparable domestic manufacturing capability.

Capital allocation and market structure differences. China's largely state-directed and state-financed energy infrastructure investment model can prioritize capacity addition and strategic infrastructure goals in ways that do not need to satisfy the same commercial return requirements, competitive bidding processes, and regulatory cost-recovery approval steps that shape U.S. utility investment decisions, particularly in the deregulated ISO/RTO markets described in Chapter 1, where merchant generation developers must secure adequate revenue certainty (often through long-term contracts, capacity market clearing prices, or corporate power purchase agreements like the AI-datacenter-nuclear deals described in Chapter 9) before committing to major capital investment — a more market-tested but often slower process than direct state investment decisions.

What "Keeping Pace" Would Require

Given this gap, what would it realistically take for the United States to meet its own future AI-driven load growth at a pace that maintains grid reliability, without attempting to simply replicate China's state-directed model, which would be inconsistent with core American governance structures and would carry its own significant trade-offs?

Most energy policy analysts, national laboratory researchers, and grid operators who have examined this question converge on a broadly similar set of priorities, most of which have already been introduced at various points throughout this book. **Permitting reform** for both transmission and generation projects — streamlining environmental review timelines without eliminating substantive environmental protections, and resolving the interstate cost-allocation disputes that have stalled major transmission projects — remains the single most frequently cited priority, and the one where the gap between current U.S. practice and what would be required to meaningfully close the pace gap with China is most significant. **Continued acceleration of the grid-enhancing technologies** described in Chapter 7 — dynamic line rating, reconductoring, and advanced power flow control — offers a comparatively fast, lower-cost way to expand effective transmission capacity without waiting for full permitting reform to be enacted. **Domestic manufacturing capacity expansion** for solar panels, wind turbine components, batteries, and, increasingly, gas turbines and nuclear reactor components, building on the incentives already established through the Inflation Reduction Act and CHIPS Act, would reduce U.S. dependence on Chinese-dominated supply chains and reduce the vulnerability of U.S. infrastructure buildout to geopolitical disruption or trade policy conflict. **Sustained, predictable federal and state policy support** across changes in political administration — a persistent challenge given the degree to which U.S. energy policy, unlike China's more continuous state planning approach, has shifted meaningfully with changes in presidential administration and control of Congress, introducing a policy uncertainty that makes long-term infrastructure investment decisions more difficult for utilities, developers, and technology companies alike. And **continued development of the emerging generation technologies** described in Part IV

— small modular reactors, enhanced geothermal, and long-duration storage — offers the prospect of new, domestically developed technology exports that could, if successfully commercialized at scale, provide the United States with genuine technological leadership in exactly the categories where China has otherwise established a commanding lead in conventional solar, wind, and battery manufacturing.

A Sober Conclusion

None of these priorities are secret or novel; each has been the subject of extensive policy discussion, legislative proposals, and regulatory reform efforts throughout the mid-2020s. What this chapter's comparison to China makes clear is the scale of urgency those efforts require. The AI-driven demand growth described throughout this book is not a distant, speculative future scenario — it is already reshaping capacity markets, reopening shuttered nuclear plants, and rewriting emergency load-shedding protocols, as earlier chapters have documented. Whether the United States can expand its generation and transmission capacity fast enough to meet that growth while maintaining the reliability Americans have long taken for granted will depend substantially on whether the structural reforms described in this chapter — permitting reform above all — can be achieved at a pace and scale that has, to date, largely eluded the American political and regulatory system, even as the urgency of the underlying challenge has become undeniable.

Case Study: Comparing Two Transmission Projects

A concrete illustration of the pace gap described in this chapter can be found by comparing specific, comparably scaled transmission projects in each country. The TransWest Express transmission line, an approximately 730-mile high-voltage direct current line designed to carry wind power from Wyoming to markets in the Desert Southwest, was first proposed in the early 2000s and did not begin construction until the early 2020s, with completion targeted for the mid-2020s — meaning the project required roughly two decades from initial proposal to completion, a timeline dominated not by the physical construction period itself (which, once permitting was complete, proceeded on a multi-year but comparatively modest timeline) but by the multi-state, multi-agency permitting and environmental review process, along with the extended negotiation required to secure transmission capacity commitments and financing.

By contrast, individual ultra-high-voltage transmission line projects in China, of comparable or greater length and capacity, have frequently moved from initial planning approval to full commercial operation within a span of a few years, a difference in overall timeline of roughly an order of magnitude compared to the TransWest Express example. This is not to suggest China's approach involved no environmental review or local consultation whatsoever — China does maintain environmental impact assessment requirements for major infrastructure projects — but the scope, duration, and multi-jurisdictional complexity of that review process differs fundamentally from the U.S. experience, reflecting the different institutional structures and political values each country's system encodes, as discussed in this chapter's main text.

The Interconnection Queue as a Microcosm of the Broader Pace Gap

The interconnection queue backlog described in Chapter 3 offers another concrete, quantifiable illustration of the broader U.S.-China pace gap examined in this chapter. The total volume of generation and storage capacity awaiting interconnection approval in U.S. queues, described in Chapter 3 as exceeding the country's entire existing installed generation fleet, represents an enormous quantity of already-planned, already-financed (in many cases) capacity that remains unable to actually begin serving load due to study and approval delays rather than any underlying technical or financial obstacle to the projects themselves. This queue backlog is, in a real sense, a symptom of the same underlying structural pace gap discussed throughout this chapter: a fragmented, multi-layered review and approval process that, whatever its other merits, has proven poorly suited to processing the sheer volume of new project proposals that falling renewable and storage technology costs have generated over the past decade.

China's more centrally coordinated planning and approval process has not faced a comparable interconnection queue backlog crisis of the scale described for the United States, in substantial part because its more unified grid planning authority can proactively plan transmission and generation capacity additions in a more coordinated, top-down fashion, rather than relying primarily on individual project developers to navigate a fragmented, developer-initiated interconnection request process of the kind that has produced the U.S. queue backlog. This comparison does not necessarily argue for wholesale adoption of China's centrally planned model — again, a model that carries its own significant trade-offs inconsistent with core American governance values — but it does illustrate concretely how specific structural choices in grid planning and interconnection processes translate into measurably different real-world outcomes in the pace at which new capacity can actually begin serving customers.

Toward a Realistic U.S. Reform Agenda

Drawing together the priorities identified in this chapter's main discussion, it is worth being specific about what a realistic, politically achievable U.S. reform agenda addressing the pace gap with China might include, distinguishing between reforms that enjoy meaningful bipartisan support and appear achievable in the near term, and more ambitious structural changes that would require a more fundamental reconsideration of the U.S. federalist regulatory model. In the more achievable category: continued federal interconnection queue reform building on

FERC's 2023 rules, expanded federal funding and technical assistance for grid-enhancing technology deployment, continued domestic manufacturing incentives for solar, battery, and nuclear supply chains, and targeted federal transmission siting authority for a narrow category of nationally significant interregional transmission projects, an authority Congress has debated extending to FERC in various permitting reform proposals without yet reaching final agreement as of the mid-2020s.

In the more ambitious, structurally significant category: a genuinely unified national transmission planning and cost-allocation framework, comparable in coordination authority (though not necessarily in ownership structure) to China's State Grid model, would likely close much of the remaining pace gap described in this chapter, but would require a degree of federal authority over historically state-regulated utility and transmission siting decisions that has, to date, faced substantial political resistance from states protective of their traditional regulatory authority over utilities operating within their borders — a resistance rooted in genuine and longstanding American federalist principles that this book does not take a position on overriding, but that readers should understand as the central political obstacle standing between the current fragmented U.S. system and the more unified, faster-moving planning model China's example demonstrates is achievable under a different governance structure.

Quantifying the Gap: A Illustrative Comparison

To make the pace comparison discussed throughout this chapter more concrete for the reader, it is useful to think through an illustrative, order-of-magnitude comparison rather than relying purely on qualitative description. Consider a hypothetical but broadly representative large-scale transmission and generation expansion program of the kind both countries have pursued in recent years: in China, a several-gigawatt UHV transmission corridor paired with associated new generation capacity at each end might reasonably proceed from initial government planning approval to full commercial operation within a window of roughly three to five years, reflecting the streamlined, centrally coordinated approval process described throughout this chapter. A comparably scaled interregional transmission and generation expansion program in the United States, by contrast, requiring coordination across multiple state jurisdictions, multiple utility service territories, and the layered federal and state environmental review processes described throughout this chapter, has more typically required a decade or more from initial proposal to full commercial operation, as illustrated by the TransWest Express example discussed earlier in this chapter.

This rough three-to-five-times difference in typical development timeline — and it is worth emphasizing this is an illustrative, order-of-magnitude comparison rather than a precise, rigorously derived statistic, since individual projects in both countries vary considerably around these typical figures — represents, in the view of most energy infrastructure researchers and policymakers who have studied this comparison, the single most consequential structural difference between the two countries' approaches to energy infrastructure development, and the one most directly relevant to this book's central question of whether the United States can build fast enough to meet its own future load growth, including but not limited to the AI-datacenter-driven growth that has been this book's primary focus.

A Note on Data Limitations and Comparative Uncertainty

Readers should approach the specific comparisons offered throughout this chapter with appropriate awareness of the data limitations involved in any U.S.-China energy infrastructure comparison. Chinese government energy statistics, while considerably more transparent and detailed than in previous decades, do not always offer the same level of independently verifiable, granular project-level detail that U.S. regulatory filings, ISO market data, and independent research organizations provide for the U.S. system, meaning some of the aggregate capacity addition and timeline figures cited throughout this Part necessarily rely on a combination of official Chinese government statistics, international energy agency analysis (including from the International Energy Agency and other organizations that track global energy infrastructure development), and independent research organizations that specialize in tracking Chinese energy sector developments, including organizations that monitor satellite imagery and other independent data sources specifically to verify or supplement official Chinese government reporting. This data limitation does not undermine the broad, qualitative conclusion this Part has emphasized — that China's pace of energy infrastructure development substantially exceeds the United States' recent pace across nearly every generation and transmission technology category discussed — but it does counsel appropriate caution regarding the precision of any specific quantitative comparison offered throughout this book, a caution consistent with the broader intellectual humility this book has emphasized, particularly in Chapter 14's closing discussion of forecasting uncertainty.

Learning from Success Stories Within the U.S. System

Despite this chapter's generally sobering comparison, it is worth noting that the U.S. system has, in specific instances, demonstrated it can move quickly when circumstances align favorably. The CREZ transmission buildout in Texas described in Chapter 4, ERCOT's comparatively fast interconnection process for new generation, and the rapid deployment of utility-scale battery storage in both California and Texas described in Chapter 5 all demonstrate that the American system is capable of the kind of pace this chapter's China comparison suggests it generally lacks, when a sufficiently aligned combination of political will, streamlined single-state regulatory jurisdiction (rather than the multi-state coordination that has slowed interregional projects), and clear economic incentive comes together. These domestic success stories offer a genuinely useful template: they suggest that the barriers to faster U.S. infrastructure deployment are less a matter of inherent national incapacity and more a matter of the specific jurisdictional fragmentation and multi-layered review processes that apply most acutely to interstate and interregional projects — precisely the category of project, including major interregional transmission lines, where the pace gap with China's unified national system is most pronounced.

Bilateral Engagement and Areas of Potential Cooperation

Despite the competitive framing that has dominated much of this Part's discussion, it is worth noting that the United States and China have, at various points, engaged in bilateral discussions touching on clean energy technology and climate cooperation, including joint statements on methane reduction and renewable deployment made at various points during the 2020s alongside the broader competitive dynamics described throughout this Part. Some energy researchers have argued that continued technical and scientific exchange between the two countries, even amid broader strategic competition, could benefit both nations' efforts to address the shared reliability and decarbonization challenges described throughout this book, particularly in areas like grid-forming inverter standards, long-duration storage research, and advanced reactor safety research where the underlying science is not inherently zero-sum. Whether such cooperation continues or narrows further amid broader geopolitical tension will likely depend on factors well beyond the energy policy scope of this book, but it is worth noting as a genuine, if currently modest, counterpoint to the primarily competitive framing that has otherwise characterized this Part's discussion of the two countries' parallel infrastructure buildouts.

PART VI
SYNTHESIS

Chapter 13:

Reliability in a Multi-Source, High-Demand Grid

The preceding twelve chapters have each examined a piece of the American grid reliability puzzle in relative isolation: traditional generation's continued importance, solar and wind's growth and operational challenges, battery storage's emergence as a shock absorber, the AI datacenter demand shock, the reordering of emergency load-shedding priorities, the promise of small modular reactors and other emerging technologies, and the sobering comparison to China's infrastructure buildout pace. This chapter brings these threads together, examining how they interact as a single system, and offers a framework for thinking about plausible future scenarios through the early-to-mid 2030s.

The Grid as a Portfolio, Not a Single Technology Bet

The most important synthesizing insight from this book is that American grid reliability in the AI era does not depend on any single technology or strategy succeeding in isolation. It depends on a portfolio of resources and strategies, each addressing a different piece of the reliability challenge, working together.

Traditional coal, gas, nuclear, and hydroelectric generation, described in Chapter 2, continue to provide the firm, weather-independent capacity the grid depends upon during its most stressed hours, even as the mix within that category shifts — coal retiring, gas expanding in the near term despite its own fuel-supply vulnerabilities exposed by Winter Storm Uri, and existing nuclear plants experiencing a genuine renaissance driven substantially by datacenter demand for carbon-free firm power. Solar and wind, described in Chapters 3 and 4, provide low-cost energy and meaningful (though not unlimited, and unevenly distributed across seasons and hours) contributions to reliability, at the cost of introducing genuine operational complexity — the duck curve, curtailment, wind droughts — that the rest of the system must be engineered to accommodate. Battery storage, described in Chapter 5, has emerged with remarkable speed as the resource best suited to smoothing out solar and wind's daily variability, while remaining, in its current four-hour-duration-dominant form, an incomplete substitute for the multi-day firm capacity traditional generation and future long-duration storage and emerging firm technologies must still provide.

Layered on top of this already-complex resource mix is the demand-side transformation described in Part III: an unprecedented surge in electricity demand concentrated in AI datacenters, met through a combination of accelerated traditional generation construction (constrained by real manufacturing bottlenecks), grid-enhancing technologies that extract more capacity from existing infrastructure, novel large-load tariffs that shift risk and curtailment responsibility onto the customers driving the growth, and a fundamental reordering of emergency operating procedures that places those same large, often self-generating-capable customers first in line for curtailment ahead of residential and most industrial customers. And looking further ahead, Part IV's emerging technologies — small modular reactors, enhanced geothermal, long-duration storage, and eventually perhaps fusion — offer the prospect of new firm, carbon-free capacity that could ease the tension between decarbonization and reliability that has defined much of this book's narrative, though on a timeline that will not meaningfully affect the reliability challenges of the remainder of the 2020s.

Three Scenarios Through 2035

It is useful to think through how these threads might plausibly combine over the coming decade, not as confident predictions but as a framework for understanding the range of outcomes that current trends make plausible.

In a best-case scenario, several favorable developments occur together. Federal and state permitting reform meaningfully accelerates transmission and generation construction timelines. Grid-enhancing technologies achieve widespread deployment, unlocking substantial additional capacity on existing infrastructure at low cost and short timelines. Gas turbine manufacturing capacity expands to meet the current order backlog, and additional gas capacity comes online roughly on the schedule utilities currently project. Battery storage costs continue falling and long-duration storage technologies (iron-air, flow batteries, and others described in Chapter 10) achieve meaningful commercial scale, closing much of the multi-day reliability gap that four-hour lithium-ion batteries alone cannot fill. Small modular reactor demonstration projects succeed both technically and economically, validating the manufacturing-based cost-reduction thesis and enabling a subsequent wave of commercial orders reaching commercial operation in the early-to-mid 2030s. AI datacenter demand growth, while still substantial, moderates somewhat from the most aggressive current forecasts as efficiency improvements in AI chips and datacenter cooling partially offset growth in computing volume, and as demand flexibility contracts allow a meaningful share of datacenter load to be served without requiring fully firm capacity commitments. In this scenario, the United States successfully navigates the demand surge with manageable cost increases and without a significant deterioration in reliability, emerging with a more diverse, more flexible, and genuinely more resilient grid than it had in the early 2020s.

In a base-case scenario, more consistent with the pace of change actually observed through the mid-2020s, progress occurs on most of these fronts but more slowly and unevenly than the best case assumes. Permitting reform remains politically contested and only partially enacted, continuing to slow major transmission and some generation projects. Gas turbine and other equipment backlogs persist for several years before manufacturing capacity catches up with demand. Battery storage continues its rapid near-term growth, but long-duration storage technologies take longer than hoped to achieve full commercial cost competitiveness. Small modular reactor demonstration projects experience some further cost and schedule setbacks, similar to the pattern seen with NuScale's Idaho project cancellation described in Chapter 9, delaying meaningful commercial-scale SMR deployment into the mid-to-late 2030s rather than the early 2030s. AI datacenter demand growth continues at a pace close to current elevated forecasts, straining capacity markets (producing further sharp capacity price increases similar to those already observed in PJM's recent auctions) and requiring more frequent reliance on the reordered emergency load-shedding hierarchy described in Chapter 8, including periodic curtailment of large flexible loads during tight conditions, though generally without triggering residential outages except during the most extreme, multi-day weather events. In this scenario, the grid remains broadly reliable but under sustained and visible strain, with electricity costs rising meaningfully faster than in the best-case scenario, and with periodic, closely watched near-miss events that keep grid reliability firmly in the public and political conversation.

In a stress-case scenario, several unfavorable developments compound each other. Permitting reform fails to advance meaningfully, and interregional transmission expansion continues to lag well behind what reliability studies indicate is needed. Extreme weather events — a severe, multi-state winter storm similar to or more severe than Uri, or an extended summer heat dome combined with a multi-day wind drought — occur during a period when reserve margins are already thin due to the combination of continued generation retirements, slower-than-hoped new firm capacity additions, and AI datacenter demand growth that continues largely unmoderated. In this scenario, the reordered load-shedding hierarchy described in Chapter 8, however well designed, proves insufficient to prevent residential and industrial customers from experiencing involuntary outages during the most extreme events, echoing the severity (though perhaps not the exact circumstances) of Winter Storm Uri. Such an event would likely trigger renewed and intense political and regulatory scrutiny of the pace of the AI datacenter buildout, potentially including more aggressive interconnection moratoria, faster-tracked but more expensive emergency generation procurement, and a broader public reckoning with the trade-offs between AI infrastructure growth and grid reliability that have, to date, been debated substantially within specialist utility, regulatory, and energy policy circles rather than in the broader public consciousness.

The Central Tension: Speed Versus Deliberation

Underlying all three scenarios is a tension this book has returned to repeatedly: the American grid's traditional planning, permitting, and construction processes were designed for an era of slow, predictable, multi-decade load growth, and those processes — for good and defensible reasons rooted in environmental protection, local community input, and careful engineering review — are inherently deliberate and slow. The AI datacenter demand surge described in Part III is arriving on a timeline that is fundamentally incompatible with that traditional pace. Something has to give: either the pace of AI infrastructure growth moderates to match the grid's traditional development timeline, or the grid's development processes accelerate to match AI's timeline, or — as appears to be happening in practice, based on the evidence assembled throughout this book — some combination of both occurs simultaneously, alongside a real, measurable increase in reliability risk and cost that the system absorbs as it works through the transition.

This is not a comfortable conclusion, but it is an honest one, and it sets up the final chapter's attempt to identify what specific developments readers should watch for as the clearest indicators of which of these paths the country is actually on.

Regional Variation in How the Story Unfolds

While this book has generally discussed the reliability challenges and responses described throughout Parts I through V at a national level, it is worth emphasizing, in this synthesizing chapter, how differently this story is likely to unfold across different regions of the country, given the significant regional variation in resource mix, regulatory structure, and datacenter growth concentration documented throughout the book. The PJM footprint, and Virginia specifically, faces perhaps the most acute near-term version of the demand-growth and reliability-strain story described in Part III, given the extraordinary concentration of existing and planned datacenter capacity in Northern Virginia, layered onto a capacity market already showing the dramatic price increases described in Chapter 1 and Chapter 6. ERCOT faces a somewhat different version of the challenge, combining substantial datacenter growth with the state's own preexisting reliability vulnerabilities exposed by Winter Storm Uri and its comparative electrical isolation from neighboring grids, though partially offset by the state's unusually rapid battery storage and renewable energy deployment described in Part II and its comparatively fast, market-friendly interconnection and permitting environment relative to many other states.

The Southeast — Georgia, the Carolinas, and neighboring states — faces substantial datacenter growth within the vertically integrated utility regulatory model described in Chapter 1, a structure that in some respects allows for more coordinated, utility-directed capacity planning (since these utilities retain direct control over generation investment decisions, rather than relying primarily on competitive wholesale markets to attract new capacity) but that also concentrates planning risk and decision-making authority more heavily within individual utility companies and their state regulatory commissions, a structure whose success in navigating the demand surge will depend substantially on the quality of individual state regulatory oversight and utility planning, evaluated on a state-by-state basis rather than through the more regionally standardized ISO/RTO market mechanisms operating elsewhere. Understanding this regional variation matters because it means the national-level scenarios described in this chapter's main discussion will likely play out unevenly, with some regions experiencing considerably more acute strain, and potentially reaching the stress-case scenario described in this chapter, well before or independent of whether the country as a whole follows a best-case, base-case, or stress-case trajectory.

The Role of Individual Consumers and Households

This book has focused substantially on the utility, regulatory, and technology-company decisions that will shape grid reliability over the coming decade, but it is worth briefly noting the role individual households and consumers can play, both in their own resilience planning and in supporting the broader transition described throughout this book. Households in regions facing elevated reliability risk, particularly during extreme winter or summer weather events, can take practical steps including maintaining basic emergency preparedness supplies, considering battery backup systems (increasingly available as a residential product, often paired with rooftop solar, and functioning as a small-scale version of the same battery storage technology described in Chapter 5), and staying informed about utility conservation appeals and demand response program enrollment opportunities, which offer households a way to both reduce their own electricity costs and contribute, in aggregate, to the virtual power plant and demand-response capacity described in Chapters 8 and 10.

More broadly, public engagement with state regulatory proceedings addressing large-load tariffs, transmission siting, and resource planning — proceedings that are generally open to public comment but that have historically attracted relatively limited public participation compared to the specialized utility, industry, and advocacy organization participants who dominate them — represents a genuine, if underutilized, avenue through which individual citizens and communities can influence the specific choices, described throughout this book, that will determine how the reliability and cost trade-offs of the AI-driven grid transition are ultimately allocated between residential customers, industrial customers, and the datacenter developers driving much of the current growth.

Revisiting the Book's Opening Question

This book opened by describing the quiet, background confidence Americans have traditionally placed in their electric grid, and the ways that confidence has been shaken by extreme weather events, rising capacity market prices, and the emergence of AI-driven demand growth. Having now traced the full arc of this book's argument — traditional generation's continued but evolving role, the renewable transition's genuine benefits and real operational trade-offs, battery storage's remarkably rapid emergence as an essential complementary technology, the unprecedented scale and speed of the AI datacenter demand surge, the quiet but consequential reordering of emergency load-shedding priorities, the promise and realistic timeline of emerging generation technologies, and the sobering comparison to China's infrastructure development pace — it is worth returning explicitly to that opening question: is the American grid becoming more or less reliable as it navigates this transition?

The honest answer this book has arrived at is that reliability is not simply increasing or decreasing in a single direction; it is being actively renegotiated and redistributed across different customer classes, different regions, and different types of risk. Residential customers, protected by the reordered load-shedding hierarchy described in Chapter

8, may in some respects face a reduced risk of routine involuntary outages during moderate capacity emergencies compared to the traditional rotating-outage system, even as the overall system operates with less comfortable reserve margins than in previous decades, and even as the risk of a genuinely severe, multi-day, wide-area event — a repeat of Winter Storm Uri at a larger scale, potentially compounded by even higher datacenter load than existed in 2021 — arguably remains as significant, or more significant, than it was before this transition began, given the continued retirement of traditional generation and the continued growth of demand described throughout this book.

The Value of Systems Thinking

If this book has a single overarching methodological lesson to offer the reader, beyond the specific facts and figures assembled throughout its chapters, it is the value of systems thinking when approaching a challenge as complex as grid reliability. It would be possible to write a much shorter, simpler account of any single thread examined in this book — a book solely about AI datacenter demand growth, or solely about battery storage, or solely about the China comparison — and such a book might offer a clearer, more digestible narrative. But it would also risk seriously misleading the reader about how the actual system works, because, as this book has emphasized repeatedly, none of these threads operates in isolation. A change in coal retirement policy affects the reserve margin available to absorb datacenter load growth. A change in battery storage economics affects how much curtailment renewable projects experience, which affects the case for demand flexibility contracts with datacenters seeking to absorb that curtailed power. A change in the pace of small modular reactor commercialization affects how quickly the tension between decarbonization commitments and datacenter firm-power requirements can be resolved. Understanding grid reliability in the 2020s and 2030s requires holding all of these interconnections in view simultaneously, exactly the approach this book has aimed to model for its readers throughout its fourteen chapters.

The Interplay Between Cost and Reliability

One synthesizing tension this book has touched upon but deserves explicit emphasis here is the relationship between cost and reliability. Nearly every strategy described throughout this book for maintaining or improving reliability amid the demand surge — new gas turbines, expanded transmission, battery storage, nuclear power purchase agreements, grid-enhancing technology deployment — carries real capital and operating costs that ultimately flow through to electricity rates paid by some combination of residential, industrial, and large-load commercial customers, depending on the cost-allocation rules discussed in Chapter 8. There is no cost-free path to the reliability this book's readers presumably want maintained; the central policy question running throughout this book's later chapters is not whether reliability will cost something, but how those costs should be allocated fairly across the customer classes that benefit from, and in some cases drive, the investment required to sustain it. Readers should carry this cost-allocation lens forward when evaluating any specific reliability proposal discussed in the news or in state regulatory proceedings, asking not only whether a given measure improves reliability, but who ultimately pays for it and whether that allocation reflects a defensible connection between cost and benefit.

A Decade-by-Decade Look Ahead

It is useful to close this synthesis chapter by sketching, in broad strokes, how the themes of this book might reasonably be expected to unfold across three successive time horizons, each corresponding to a different mix of the strategies and technologies discussed throughout the preceding chapters.

Through the late 2020s, the grid's response to AI-driven demand growth will be dominated overwhelmingly by the near-term tools described in Chapter 7: accelerated natural gas construction working through existing turbine backlogs, grid-enhancing technology deployment on existing transmission infrastructure, continued rapid battery storage buildout, existing nuclear plant life extensions and uprates, and the large-load tariff and curtailment reforms described in Chapter 8 becoming standard practice across an increasing number of states. Capacity markets in datacenter-heavy regions will likely remain under sustained upward price pressure during this period, and the reordered emergency load-shedding hierarchy will be tested, in a modest way, during ordinary seasonal peak-demand events, well before it faces the kind of severe, multi-day test a future extreme weather event would provide.

Through the early-to-mid 2030s, the first wave of small modular reactor and enhanced geothermal projects described in Part IV should reach commercial operation, assuming the demonstration projects currently underway succeed on a timeline reasonably close to current announcements, providing the first genuinely new source of firm, carbon-free capacity specifically developed in response to the AI datacenter demand surge described throughout this book. Long-duration storage technologies should begin reaching meaningful commercial scale during this period as well, narrowing the gap between four-hour lithium-ion battery capability and the multi-day firm capacity traditional generation has historically provided. Whether federal and state permitting reform has meaningfully accelerated transmission development by this point will likely determine whether the interregional transmission constraints described throughout Parts II and III have eased or continued to bind.

Beyond the mid-2030s, assuming continued progress along the lines described above, the American grid could plausibly emerge as a genuinely more diversified, more flexible, and more resilient system than the one it inherited at the start of this transition – combining a smaller but still meaningful traditional generation fleet, a much larger renewable and storage base, a growing contribution from small modular reactors and other emerging firm technologies, and a demand side that has itself become considerably more flexible through datacenter curtailment arrangements, managed electric vehicle charging, and virtual power plant aggregation of distributed resources. This is, to be clear, the more optimistic end of the range of outcomes this chapter's three scenarios describe, and readers should hold it as a plausible destination rather than a guaranteed one, given the very real risks of delay, cost overrun, and extreme weather disruption discussed throughout the base-case and stress-case scenarios earlier in this chapter.

Chapter 14:

Conclusion — What Reliability Will Mean in 2035

This book opened with an observation about how quiet and boring grid reliability used to be – a background assumption so reliable that most Americans never thought about it at all. It closes with the recognition that this quiet confidence is unlikely to fully return, at least not in the form it took in the twentieth century. The grid of 2035 will be a more complicated, more actively managed, and in many ways more interesting system than the grid Americans grew up with – but whether it is also a more reliable one is a genuinely open question, and one this final chapter attempts to answer as honestly as the evidence allows.

Key Takeaways

For the policymaker, several conclusions from this book deserve particular emphasis. Permitting reform for transmission and generation infrastructure is not a peripheral or purely partisan issue; it is arguably the single highest-leverage lever available to ensure the United States can meet AI-driven demand growth without unacceptable reliability or cost consequences, and the comparison to China's infrastructure buildout pace in Part V makes clear how much ground the current U.S. system has to make up. Large-load interconnection tariffs and the reordered emergency load-shedding hierarchy described in Chapter 8 represent a genuinely constructive innovation – protecting residential customers by placing curtailment responsibility on the large, often self-generating-capable customers driving new demand – and deserve continued refinement and, where not yet adopted, broader implementation across additional states and utilities. Policy support for both near-term solutions (grid-enhancing technologies, demand flexibility contracts, existing nuclear plant life extension) and longer-term technologies (small modular reactors, enhanced geothermal, long-duration storage) should be pursued in parallel, since no single technology timeline aligns perfectly with the urgency of the current demand growth.

For the utility planner, this book's central lesson is that the old playbook — decade-long load forecasts built on assumptions of steady, low, predictable growth — no longer applies, at least not in regions experiencing significant datacenter development. Planning processes need to accommodate genuine uncertainty about the ultimate scale of AI-driven demand, verified through the kind of financial-commitment and site-control requirements increasingly built into large-load tariffs, while building enough flexibility into resource and transmission plans to adapt as that uncertainty resolves in either direction over the coming years.

For the general reader, the most important takeaway may simply be an updated mental model: the electric grid is no longer, and will not again soon be, the background, unchanging utility it once was. It is undergoing genuine, consequential change, driven by decarbonization, generation retirements, and unprecedented AI-driven demand growth, and the choices made by utilities, regulators, and technology companies over the next several years will shape both the cost Americans pay for electricity and the reliability they can expect from it for a generation to come.

What to Watch

Several specific, observable indicators will help readers track which of the scenarios described in the previous chapter is actually unfolding, and are worth watching in the news and in utility and regulatory filings over the coming years.

Capacity market clearing prices, particularly in PJM's annual auctions, offer perhaps the single clearest, most quantifiable real-time signal of how tightly supply and demand are balancing across the country's largest wholesale market. Continued sharp increases would indicate the base-case or stress-case scenarios described in Chapter 13 are unfolding; a stabilization or moderation in clearing prices would suggest new capacity additions and demand-side measures are successfully catching up with load growth.

NERC's annual Long-Term Reliability Assessment will continue to serve as the industry's own authoritative gauge of regional reliability risk, and its year-over-year changes in the geographic scope and severity of "elevated" and "high" risk designations will indicate whether the combination of strategies described in this book is succeeding in keeping pace with demand growth and generation retirements.

Gas turbine order backlogs and delivery timelines, along with progress on the small modular reactor demonstration projects described in Chapter 9 (particularly whether TerraPower's Wyoming project, X-energy's and Kairos Power's respective partnerships with Amazon and Google, and any subsequent SMR projects meet their announced schedules and cost targets), will indicate whether the supply-side response to demand growth is proceeding on the optimistic or more delayed timeline described in the scenarios of the previous chapter.

Federal and state permitting reform legislation, and the pace of major interregional transmission project approvals specifically, will indicate whether the single highest-leverage structural reform identified in this book is actually being pursued with the urgency the situation warrants, or whether it remains, as it has for much of the mid-2020s, a subject of extensive discussion without commensurate legislative action.

AI datacenter demand growth realization versus forecast, tracked through utility interconnection queue data, actual (as opposed to announced) datacenter construction activity, and the take-or-pay commitments described in Chapter 7, will help clarify whether the most aggressive current demand forecasts are materializing in practice, moderating due to AI efficiency improvements or a cooling of speculative investment, or in fact proving conservative relative to even faster actual growth.

The frequency and severity of extreme weather events, and how the grid — including the reordered load-shedding hierarchy described in Chapter 8 — performs during them, will ultimately provide the most direct, real-world test of whether the reliability strategies described throughout this book are working as intended, in the same way Winter Storm Uri provided an unambiguous, tragic test of the pre-existing system's vulnerabilities in 2021.

Closing Thought

The American electric grid has reinvented itself before — most dramatically during the rural electrification and interstate interconnection buildout of the mid-twentieth century, and again during the deregulation and market restructuring of the 1990s and 2000s. Each of those transformations was driven by a combination of technological change, economic pressure, and policy choice, and each ultimately produced a grid that, whatever its remaining imperfections, was more capable than the one it replaced. The transformation described in this book — driven by decarbonization, generation retirement, and an unprecedented AI-driven demand surge, and addressed through battery storage, reordered emergency protocols, emerging nuclear technology, and a sobering comparison to China's infrastructure pace — is, by the evidence assembled here, already well underway. Whether it concludes as a success story of adaptation, comparable to those earlier transformations, or as a more difficult and costly period of strain and periodic crisis, will depend on choices being made right now, in places far less visible to most Americans than their own light switches: utility planning departments, state regulatory commissions, congressional committee rooms, and corporate boardrooms weighing the electricity demands of an AI-powered future against the grid's capacity to deliver it reliably. This book has aimed to make those choices, and their stakes, legible to the reader who flips the switch and, for now, still expects the light to come on.

A Final Word on Uncertainty and Humility

This book has made extensive use of forecasts, scenarios, and projections — load growth forecasts, capacity market trends, technology deployment timelines — and it is worth closing with an explicit acknowledgment of the humility such forecasting demands. Every major inflection point described in this book was, at some point in the recent past, itself a surprising deviation from previous expectations: the shale gas revolution that undercut coal's economics was not widely anticipated a decade before it occurred; the speed of solar and battery cost declines exceeded most forecasters' expectations throughout the 2010s; and the scale and suddenness of AI-driven datacenter demand growth, the central subject of this book, caught utility planning departments across the country by surprise as recently as the early 2020s, despite those same departments' decades of collective forecasting experience and sophisticated analytical tools.

This history of forecasting surprise should inform how readers approach every specific number, timeline, and scenario presented throughout this book. The broad structural forces described here — continued reliance on traditional generation for firm capacity, the operational trade-offs renewable energy introduces, batteries' emergence as an essential complementary technology, the scale and novelty of AI-driven demand growth, the reordering of emergency protocols to protect residential customers, the promise and limits of emerging generation technology, and the sobering comparison to China's infrastructure pace — are well-documented and likely to remain relevant organizing themes for understanding American grid reliability for years to come. But the specific pace, magnitude, and timeline of each development remains subject to the same kind of forecasting uncertainty that has characterized every previous chapter of the grid's ongoing evolution, and readers should hold the specific figures and projected timelines presented throughout this book with appropriate humility, updating their understanding as new information emerges in the years ahead, exactly as this book's own analysis will need to be updated and revised as the story it describes continues to unfold.

Acknowledging What This Book Has Not Covered

In the interest of intellectual honesty, it is worth briefly noting several important, related topics this book has necessarily left out or addressed only in passing, given its specific focus on the intersection of traditional generation, renewable energy, battery storage, AI-driven demand growth, emergency load-shedding protocols, emerging nuclear technology, and the U.S.-China comparison. This book has not addressed in depth the retail electricity rate design questions that determine how the wholesale market and infrastructure cost dynamics described throughout this book ultimately translate into the bills individual residential and business customers actually pay, a subject deserving its own dedicated treatment given its direct relevance to public affordability concerns. It has similarly not addressed in depth the broader climate policy and emissions trajectory questions that intersect with, but are analytically distinct from, the reliability-focused questions this book has emphasized, nor has it addressed the cybersecurity and physical security considerations that represent an entirely separate, and increasingly significant, dimension of grid reliability planning beyond the resource adequacy and capacity questions this book has focused upon. Readers interested in these related topics are encouraged to pursue them through the dedicated literature that exists on each, understanding this book as a focused treatment of one particular, currently urgent slice of the broader American grid reliability picture, rather than a comprehensive treatment of every dimension of that picture.

The story this book has told is, by its nature, a story still being written. The specific developments readers should watch for, identified earlier in this chapter, will continue to unfold in utility filings, regulatory dockets, capacity market auctions, and, unfortunately, in the real-world stress tests that severe weather events continue to provide. This book has aimed to equip its readers with the conceptual framework and factual grounding needed to follow that unfolding story with genuine understanding, rather than merely passive alarm or complacency — an understanding that, as this book has argued throughout, has never been more necessary for engaged citizens, policymakers, and utility professionals alike than it is at this genuinely pivotal moment in the history of the American electric grid.

Advice for Different Readers Going Forward

Different readers of this book will carry away different practical implications, and it is worth closing with a few direct words to each. If you are a utility or grid-operator professional, this book's account should reinforce the case for the portfolio approach described in Chapter 13: continued investment in traditional firm generation life-extension and new gas capacity where needed, aggressive deployment of grid-enhancing technologies and battery storage, rigorous large-load interconnection tariff design that protects existing customers while still accommodating legitimate new growth, and active engagement with the emerging generation technologies described in Part IV, even though their full commercial impact remains a decade or more away. If you are a policymaker or regulator, this book's account should reinforce the urgency of permitting reform, both for transmission and for the generation technologies needed to replace retiring capacity, and should encourage continued attention to the cost-allocation and consumer-protection questions that determine whether the benefits and burdens of the AI-driven grid transition are distributed fairly across residential, industrial, and large-load commercial customers. If you are a technology industry professional working on AI infrastructure development, this book's account should underscore both the genuine electricity supply constraints your industry's growth is creating for the broader grid, and the meaningful opportunities — demand flexibility, direct investment in nuclear and geothermal generation, and transparent, well-structured large-load interconnection agreements — through which your industry can be a constructive partner in addressing those constraints rather than simply a source of strain upon the system.

And if you are simply a general reader who picked up this book because you had noticed grid reliability becoming a more frequent subject of news coverage and wanted to understand why, this book's account should leave you with a clearer sense of the specific forces at work — decarbonization, generation retirement, and AI-driven demand growth — and the specific tools, from battery storage to reordered emergency protocols to emerging nuclear technology, that utilities, regulators, and technology companies are deploying in response. The electric grid that flips on with a light switch in your home is undergoing one of the most significant transformations in its history, arriving at a pace that will test the American energy system's institutions and infrastructure development processes as much as its underlying technology. Understanding that transformation, as this book has aimed to help you do, is the first step toward being an informed participant in the choices that will determine its outcome, whether as a voter, a ratepayer, a professional working somewhere within this vast system, or simply a citizen who, like the author, still expects the light to come on when the switch is flipped, and who now has a fuller understanding of everything required to keep that expectation a reasonable one to hold.

A Concluding Image

Consider once more the simple act with which this book began: flipping a switch and trusting that light will appear. Behind that simple act now stands an extraordinarily complex, rapidly evolving system: aging coal plants running longer than planned to preserve reliability margin; gas turbines backlogged years into the future; solar farms and wind turbines feeding an ever-larger share of the grid's energy while batteries absorb their surplus and smooth their gaps; a handful of technology companies financing the restart of shuttered nuclear reactors to feed the computing clusters training the next generation of artificial intelligence; utility lawyers and regulators quietly rewriting the rules for whose power gets cut first when the system is pushed to its limits; engineers in Wyoming and Tennessee racing to prove that nuclear reactors can be built like factory products rather than one-of-a-kind monuments; and, half a world away, Chinese state planners connecting deserts full of solar panels to coastal megacities through transmission lines longer and faster-built than anything in American history. All of it, ultimately, in service of that same simple act: light appearing when the switch is flipped. This book has tried to make that hidden complexity visible, and in doing so, to help its readers understand not just how the grid works today, but how the choices being made right now will determine whether it keeps working, reliably, for the demanding new economy the country is building on top of it.

Gratitude and Continued Inquiry

Writing this book required synthesizing material from utility regulatory filings, grid operator technical reports, corporate announcements, and journalistic accounts spanning electrical engineering, energy economics, corporate strategy, and international relations — a genuinely interdisciplinary undertaking that reflects the interdisciplinary nature of the grid reliability challenge itself. The author is grateful to the utility engineers, grid operators, regulators, and energy journalists whose ongoing public work made this synthesis possible, and hopes this book, in turn, provides a useful foundation for readers who wish to follow this story as it continues to develop, whether through the specific indicators identified in this chapter's "What to Watch" section, through continued independent reading in the specialized energy trade press, or through direct engagement with the state regulatory proceedings and utility planning processes where many of the specific decisions shaping this book's subject matter are actually made.

BACK MATTER

Glossary

Ancillary services – Grid support services beyond basic energy delivery, including frequency regulation, spinning and non-spinning reserves, voltage support, and black start capability, needed to keep the grid stable second-to-second.

Balancing authority – The entity responsible for maintaining real-time supply-demand balance within a defined electrical footprint.

BESS (Battery Energy Storage System) – A utility-scale installation of batteries, almost always lithium-ion as of the mid-2020s, used to store and later discharge electricity for grid services.

Capacity market – A wholesale market mechanism, used by PJM and other ISOs/RTOs, in which generation, storage, and demand-response resources are compensated for committing to be available during a future delivery year, distinct from payments for energy actually produced.

Curtailment – The deliberate reduction of a generator's (typically solar or wind) output below its available potential, due to insufficient demand, transmission constraints, or system stability requirements.

Demand response – Voluntary, pre-arranged, compensated reduction of electricity consumption by customers in response to grid conditions, activated before involuntary load shedding.

Duck curve – The characteristic shape of a net-demand graph in a high-solar-penetration grid, showing a steep midday dip (from abundant solar output) followed by a steep evening ramp as solar output falls while demand rises.

Effective load-carrying capability (ELCC) – A metric estimating how much a generation or storage resource actually reduces the probability of a supply shortfall during system-stress hours, as opposed to simply counting its full nameplate capacity.

Enhanced geothermal systems (EGS) — Geothermal power technology that artificially creates the underground conditions (heat, water, permeable rock) needed for geothermal generation at sites lacking naturally occurring geothermal resources, using techniques adapted from oil and gas drilling.

Grid-enhancing technologies (GETs) — A category of technologies, including dynamic line rating, reconductoring, and advanced power flow control, that increase the capacity of existing transmission infrastructure without new construction.

Grid-forming inverter — An advanced inverter control technology, used primarily in battery storage, capable of actively establishing and stabilizing grid voltage and frequency, as opposed to simply following conditions set by other resources.

Inertia — The physical resistance to sudden frequency change provided by the rotating mass of synchronous generators (coal, gas, nuclear, hydro), historically a "free" stabilizing byproduct of traditional generation that inverter-based renewable and storage resources do not inherently provide.

Interconnection queue — The sequential process by which new generation, storage, or large-load projects are studied and approved for connection to the grid; a major bottleneck for renewable and storage deployment as of the mid-2020s.

ISO/RTO (Independent System Operator / Regional Transmission Organization) — Regional entities that operate wholesale electricity markets and coordinate transmission planning across multiple states and utility territories, including PJM, ERCOT, MISO, CAISO, and SPP.

Large-load tariff – A utility rate structure specifically designed for very large individual customers (chiefly datacenters), often including minimum-take commitments, longer contract terms, and curtailability provisions.

Load shedding – The deliberate, controlled disconnection of customers from the grid to restore supply-demand balance during a capacity emergency, typically implemented as a last resort after conservation appeals, demand response, and voltage reduction have proven insufficient.

NERC (North American Electric Reliability Corporation) – The federally designated organization responsible for developing and enforcing mandatory reliability standards across the U.S. and Canadian bulk power system.

Reserve margin – Generating capacity maintained beyond forecasted peak demand, expressed as a percentage, sized to meet probabilistic reliability standards.

Small modular reactor (SMR) – A nuclear reactor design, generally 50 to 300 megawatts of capacity, intended to be manufactured substantially in factories using standardized, repeatable processes rather than built as a unique, site-specific megaproject.

Ultra-high-voltage (UHV) transmission – Very high voltage transmission lines, used extensively in China, designed to move large quantities of electricity over very long distances with minimal losses.

Timeline of Major Grid Events, 2020–2026

2020 — California experiences rotating blackouts during an August heat wave, driven by a regional heat event reducing imported power availability alongside high in-state demand.

February 2021 — Winter Storm Uri causes catastrophic, multi-day generation failures across Texas, driven substantially by frozen natural gas infrastructure; ERCOT implements emergency rotating outages; more than 200 deaths are attributed to the event; sweeping weatherization reforms follow.

2022 — The Inflation Reduction Act is enacted, extending and expanding federal tax credits for solar, wind, storage, and nuclear generation, and establishing new domestic manufacturing incentives; large AI language models begin attracting mainstream commercial and public attention, triggering the first wave of major datacenter capacity announcements.

2023 — NuScale Power's planned Idaho SMR project is canceled due to cost increases and insufficient subscriber commitment; the Federal Energy Regulatory Commission issues major interconnection queue reform rules; utility long-term load forecasts across multiple states begin showing significant upward revisions attributed to datacenter growth.

2024 — Microsoft and Constellation Energy announce an agreement to restart a unit at the former Three Mile Island site (rebranded the Crane Clean Energy Center) to supply datacenter power; Amazon announces agreements with Talen Energy (Susquehanna nuclear plant) and X-energy (small modular reactor development); Google announces an agreement with Kairos Power for SMR development.

2024–2025 – PJM's capacity market auctions clear at dramatically higher prices than prior years, cited by PJM and independent market monitors as reflecting generator retirements, slower new capacity additions, and datacenter-driven load growth; multiple utilities announce large-load interconnection tariff proposals to state regulators.

2025 – Multiple states (including Georgia, Ohio, Virginia, and Indiana) advance regulatory proceedings addressing large-load interconnection terms and curtailment provisions specifically targeting datacenter customers; gas turbine order backlogs stretch to the end of the decade at major manufacturers.

2026 – Continued construction progress (or delay) on TerraPower's Wyoming Natrium demonstration project and other SMR developments; continued growth in utility-scale battery storage deployment in California, Texas, and other markets; ongoing national debate over federal permitting reform legislation for transmission and generation infrastructure.

Notes on Sources

This book synthesizes publicly available information from multiple categories of sources current through 2026: annual and long-term reliability assessments published by the North American Electric Reliability Corporation (NERC); market and operational reports published by PJM Interconnection, ERCOT, CAISO, MISO, and SPP; federal agency publications and rulemakings from the Federal Energy Regulatory Commission, the Department of Energy, and the Nuclear Regulatory Commission; corporate announcements and regulatory filings from utilities including Dominion Energy, Georgia Power, Duke Energy, American Electric Power, Constellation Energy, and Talen Energy; corporate announcements from technology companies including Microsoft, Amazon, and Google regarding datacenter energy procurement; research and analysis from national laboratories including Lawrence Berkeley National Laboratory and the National Renewable Energy Laboratory; and reporting and analysis from established energy industry trade publications and mainstream business and technology press.

Given the rapidly evolving nature of the subject matter — particularly AI datacenter demand forecasts, individual project timelines, and market clearing prices — readers seeking the most current figures on any specific data point discussed in this book (capacity market clearing prices, specific project completion dates, updated load forecasts) are encouraged to consult the primary sources named above directly, as figures presented here reflect the most reliable information available at time of writing and are, in a fast-moving sector, subject to ongoing revision.

Index

Frequently Referenced Organizations

NERC (North American Electric Reliability Corporation) – The federally designated reliability standards organization for the U.S. and Canadian bulk power system, publisher of the annual Long-Term Reliability Assessment referenced throughout this book.

FERC (Federal Energy Regulatory Commission) – The federal agency regulating interstate electricity transmission and wholesale markets, and the approving authority for ISO/RTO market rules and interconnection reforms discussed throughout Parts II and III.

PJM Interconnection, ERCOT, MISO, CAISO, SPP – The major U.S. Independent System Operators and Regional Transmission Organizations discussed throughout this book, each responsible for wholesale market operation and reliability coordination within its footprint, as introduced in Chapter 1.

Nuclear Regulatory Commission (NRC) – The federal agency responsible for licensing and safety regulation of U.S. commercial nuclear reactors, including the small modular reactor designs discussed in Chapter 9.

Department of Energy (DOE) – The federal cabinet department responsible for a wide range of energy policy, research funding, and loan guarantee programs referenced throughout Parts III and IV, including support for grid-enhancing technologies, domestic manufacturing, and advanced nuclear demonstration projects.

State Grid Corporation of China and China Southern Power Grid – The two dominant state-owned grid operating companies in China, discussed in Chapter 11, whose unified planning authority contrasts with the fragmented U.S. ISO/RTO and utility regulatory landscape.

A Reader's Companion Checklist

For readers who wish to track the developments discussed throughout this book as they continue to unfold, the following checklist, organized by the "What to Watch" indicators introduced in Chapter 14, may prove useful for ongoing monitoring:

- Annual PJM (and other ISO) capacity market auction results and clearing prices
- NERC's annual Long-Term Reliability Assessment and its regional risk designations
- Gas turbine and large power transformer order backlogs and delivery timelines reported by major manufacturers
- Progress updates on TerraPower's Wyoming Natrium project, X-energy's and Kairos Power's respective datacenter-linked SMR partnerships, and any newly announced advanced reactor projects
- Federal permitting reform legislative activity and major interregional transmission project approvals
- Utility integrated resource plans and long-term load forecasts in datacenter-heavy states, particularly Virginia, Georgia, Texas, and Ohio
- State regulatory proceedings addressing large-load interconnection tariffs and curtailment provisions
- Battery storage deployment statistics and long-duration storage technology commercialization milestones

This checklist is not exhaustive, but it captures the specific, trackable indicators this book has identified as most likely to reveal, over the coming several years, which of the scenarios described in Chapter 13 – best case, base case, or stress case — is actually unfolding across the American grid.

Author's Note

This book was written as a synthesis of publicly available information current through 2026, intended to make a genuinely complex and rapidly evolving subject accessible to the general reader and engaged undergraduate audience without requiring specialized prior training in electrical engineering or energy economics. Given the pace of change in every domain this book covers — AI technology and its electricity demand implications, battery and nuclear technology commercialization, capacity market dynamics, and the ongoing U.S. and Chinese infrastructure buildouts — readers should treat this book as a snapshot and a framework for understanding an unfolding story, rather than a final or unchanging account. The author welcomes the prospect of revisiting and updating this analysis as the story continues to develop in the years ahead.